Beyond Trawlertown

Studies in Port and Maritime History 2

Studies in Port and Maritime History

Series editors:
Andrew Popp (Copenhagen Business School)
Nicholas J. White (Liverpool John Moores University)

Placing port and maritime history together, this series is an invitation to cross boundaries and borders, publishing research of the highest quality in the fields of maritime and port history. The centrality of maritime activity to human history is long established, but all maritime activities depend on and connect ports, their hinterlands, and their communities, serving to link peoples, economies, and cultures. Ports are at the centre of regional and international trade and business networks, just as all navies, merchant or military, are dependent on ports, their services, and their infrastructures. At the same time, ports are bridge locations, focal points in processes of exchange and transfer that are social, cultural, and demographic as much as they are economic. These functions render ports unique and distinctive spaces, acting as key nodes in successive waves of globalisation. It is the sea, meanwhile, which makes all of these connections possible.

Beyond Trawlertown

Memory, Life and Legacy
in the Wake of the Cod Wars

Jo Byrne

LIVERPOOL UNIVERSITY PRESS

First published 2022 by
Liverpool University Press
4 Cambridge Street
Liverpool
L69 7ZU

British Library Cataloguing-in-Publication data
A British Library CIP record is available

ISBN 978-1-80085-655-4 cased

Typeset by Carnegie Book Production, Lancaster
Printed and bound by CPI Group (UK) Ltd, Croydon CR0 4YY

For Dad

1933–2015

Contents

Maps

Figures

Acknowledgements

I would like to extend my thanks to everyone who has helped me on this journey. I owe much to my PhD supervisors, Professor David J. Starkey, Professor David Atkinson and Robin Diaper, Curator of Maritime and Social History at Hull Museums, for their guidance and advice. A debt also goes to my colleagues at Blaydes Maritime Centre at the University of Hull, and in particular to Dr Martin Wilcox and Dr Robb Robinson, whose scholarship has helped to frame my own contribution. The research and its subsequent translation into a book was made possible with funding from the Arts and Humanities Research Council, Hull Museums and the Wilson Family Research Fellowship in Maritime History. My appreciation goes to Alison Welsby and Liverpool University Press for guiding me through the stages of publication.

A special heartfelt thank you goes to my husband Fintan for his patience, for keeping things together at home and for rescuing me from numerous 'crises' on the long road from research to authorship. Thanks also go to my daughters Eilis and Demi for giving me space to work; to my mum and my friends who have offered moral support and proofreading and to my dad for prolonged conversations about trawling.

There are so many people who have made generous contributions in bringing this story to life. It is a privilege to have permission to include the vivid and evocative documentary photography of Alec Gill, Donald Innes and Paul Hurst. I am indebted also to staff at the Butterflies Dementia Support Group, Edinburgh Street Community Centre, Hessle Road Network, Hull Museums, the Fishermen's Mission, and also to Alan Hopper, for introducing me to interviewees. And finally, the most enormous thank you goes to the narrators themselves, who shared their time, insight and memories and without whom this work could not exist. This book is dedicated to you.

Abbreviations

ABP	Associated British Ports
AHD	Authorised heritage discourse
BFF	British Fishing Federation
CFP	Common Fisheries Policy
EEC	European Economic Community
EEZ	Exclusive Economic Zone
FN	*Fishing News*
FNI	*Fishing News International*
HC	Hamling Collection
HCA	Hull City Archives
HFTOC	Humber Freezer Trawler Owners Company
HFVOA	Hull Fishing Vessel Owners Association
NAFHA	North Atlantic Fisheries History Association
NEAFC	North East Atlantic Fisheries Commission
SFIA	Sea Fish Industry Authority
TAC	Total Allowable Catch
WF	*World Fishing*
WFA	White Fish Authority

Glossary of Fishing and Nautical Terms

General

Barrow boy A youth employed to move fish around the quayside, often a first step towards gaining employment on the dock or at sea.

Bobber A Hull term for dockworkers who unloaded fish from trawlers (known in Grimsby as lumpers).

Bullnose A rounded area of dockland at the entrance to Hull's fish dock, south of the lock gates and facing the River Humber.

Demersal fish Fish living predominantly at or near the bottom of a body of water, such as cod.

Dutch auction A means of auction (used at Hull's fish market) where the auctioneer starts with a high price, which drops until a participant is willing to accept the offer.

Flag of convenience The practice of registering a merchant ship in a country different from that of the ship's owners and flying the flag of the country of registration.

Frozen at sea (FAS) Fish that is frozen aboard a trawler soon after being caught.

Green hand Nautical slang for an inexperienced crew member.

Gunboat A lightly armed naval vessel used by the Icelandic Coastguard.

Handlining A form of inshore fishing that uses baited hooks attached to lines towed by small boats.

Jigging A form of handlining that uses an unbaited lure or 'jig', which creates a vibration in the water.

Kit A local term for the dockside containers that held wet fish being sold at auction.

Klondyking A term used for the system whereby fish is transferred at sea from the catching vessel to a larger factory ship for processing.

Pelagic fish Fish that live in open water, away from the seabed or shore, such as mackerel.

Poundage A local term for the money raised from the sale of the catch, paid to the trawling crew as a percentage share to supplement their basic wage.

Quota An allocation of a share of fish stock that may be caught by a nation, company or vessel, introduced in an attempt to manage overfishing.

Runner The term used in the British trawl industry for the person employed by trawling companies with responsibility for recruiting and managing crews.

Smoke house A building for smoking fish. In Hull these were generally brick built with a steeply pitched roof and distinctive rows of tall, slim chimneys.

Stand A workspace with a small office above, located to the rear of the fish market sheds and occupied by fish merchants.

Swallow the anchor Nautical slang meaning to give up going to sea.

Transhipment In maritime terms, the off-loading of a cargo from one ship to be loaded onto another without the cargo leaving the port. In modern parlance, klondyking may be referred to as transhipment at sea.

Trawlers

Factory ship A large vessel with facilities for processing and freezing fish whilst still at sea.

Fishing smack A sail-rigged fishing vessel.

Freezer trawler Usually a stern trawler, with facilities to freeze fish at sea. Cleaning, sorting or gutting may take place before freezing, but full processing tends to be reserved for factory ships. Aboard a freezer vessel (non-factory), fish is typically frozen into blocks and the fish defrosted and processed ashore.

Japanese squid jigger A specialist vessel using lights to attract squid and 'jigging' thousands of hooks from separate winches. Factory vessels process the squid on board in large quantities.

Mothership A vessel that provides support and services to a fleet at sea. In fishing terms, it may also be a large factory or transport vessel (reefer) that receives fish from smaller vessels via the systems of klondyking or transhipment at sea.

Purse seiner (seine netter) A vessel rigged for pelagic fishing where a weighted net is released to encircle a fish shoal and drawn in like a purse string. On larger vessels fish may be pumped aboard into a brine-filled tank for freezing.

Sidewinder or side trawler The dominant trawler of the British North Atlantic distant-water industry until the late 1960s, gradually replaced with the stern trawler. Hull sidewinders were wet fish trawlers, with the trawl set over the side and the winch set forward. Fish were gutted on the open deck and stored below.

Stern trawler A major advance from the side trawler, with the trawl released and hauled from the rear, via a ramp on larger vessels. Fish is generally handled and stored below deck. Stern trawlers can be wet fish, freezer or factory vessels.

Trawls

Bottom/demersal trawl A cone-shaped net towed across the bottom of the seabed. Fish are channelled into the point of the cone, known as the cod end. The Hull industry traditionally favoured the otter trawl, where the net is held open by two large metal otter boards or 'trawl doors'.

Mid-water/pelagic trawl A cone-shaped net with cod end, designed, rigged and weighted to fish mid-water. Generally larger than a bottom trawl, it can catch fish close to the surface or at great depth. The mid-water trawl is designed for pelagic fishing, but can also target demersal species.

Purse seine net A large wall of netting, with floats on the surface and a lead line threaded around the bottom. The net encircles the shoal and the lead line is pulled to 'purse' the net and entrap the fish.

Glossary of Vernacular Terms

Aren't	Am not (used with first person singular)
Cos	Because
'em	Them
Gonna	Going to
Kid	A man, usually a younger man – or can mean child
Me	My (i.e. me dad)
Nowt	Nothing
o'	Of (i.e. a cup o' tea)
Sez	Says/said
Summat	Something
The	'The' can be omitted (i.e. I'm going to shops, going to park)
'uns	Ones (i.e. little 'uns (children))
Went	Went can mean said (i.e. He went, 'I'm going home')
Won't	With a short 'o' this means 'was not', with a long 'o' it means will not
Y'	You

Preface: Starting the Journey

Beyond Trawlertown is the product of four years of research undertaken for my PhD between 2011 and 2015. The University of Hull and Hull Maritime Museum, funded by the Arts and Humanities Research Council, had issued the invitation to consider the social, spatial and cultural aftermath of the Anglo-Icelandic fishing conflicts, known as the Cod Wars, through a case study of the trawling port of Hull. The opportunity was attractive. My own interest in British distant-water trawling was already established. My father had been a Hull trawlerman, who had come ashore in 1964 and trained fishermen until the 1980s. I was intrigued by his tales of an extreme life at sea. I had read local trawling histories and had been absorbed by Rupert Creed's oral history play *The Northern Trawl*. Most narratives had ended in the mid-1970s, with the demise of Hull as a trawling port. In written and popular accounts much less was said about what had happened next.

My family lived just beyond Hull's fishing district. I remember visiting the fish dock once as a child with my father, but can only conjure small details. My secondary school was located in the streets to the north of Hessle Road – I had friends who lived there and would shop there sometimes on a weekend. In the early 1980s, cycling to school from my grandmother's house meant inhaling the smell of batter from the Birds Eye frozen fish factory, and travelling past the boarded terraces of the later phases of housing demolition. The stench of fish meal might occasionally waft through an open classroom window, but by this time trawling had little lingering influence on the teen culture of Amy Johnson High. Yet these fleeting personal encounters have instilled the sense of inquiry and personal heritage that has driven my research and made this history the one that I wanted to tell.

The narrative that has emerged sets an emphasis on *beyond*. By taking an end as the beginning, it considers the aftermath of a time when rising oil prices, falling fish stocks and international fishing policy converged to force British distant-water trawling into rapid contraction and decline. By contemplating the subject from a cultural perspective, the research exceeds the prevailing economic, political, technical or environmental emphasis of many maritime histories. Using the oral history method, the book puts people and place at the core. Analysis is underpinned by a phenomenological attitude – one which acknowledges the influences of individual and collective consciousness, composure and cultural discourse within remembered accounts. The work is interdisciplinary, extending beyond historical analysis to embrace perspectives from cultural geography and critical heritage studies. In doing so it brings history into the present and beyond by positioning the transition of Hull's fishery from maritime industry to maritime heritage as the final but ongoing chapter.

Beyond Trawlertown is the outcome of 41 research interviews that I conducted with 43 individuals. Each narrator willingly shared their time, memories and perspectives and I am indebted to their generosity. The interviewees represent a range of positions and ranks from within Hull's fishing industry and community. Their relationships to the fishery are either direct or through the experience of a close relative. A narrator's field of knowledge may cross several areas, offering multiple insights within a single interview. Margaret, for example, was concurrently the daughter of a fish dock bobber, sister of a fisherman, a long-term Hessle Road resident, one-time fish processor and barmaid in a Hessle Road trawlermen's pub. As a result she could narrate from diverse perspectives. The instability of the research period meant that other narrators had shifted from one role to another, for example from Arctic fishing, to 'mackerelling' and then to the North Sea gas and oil sector.

The interview sample consists of 35 male and eight female narrators, reflecting the strong masculine character of the industry. Distant-water trawling was and remains a highly gendered occupation and this is evident in workplace, community and family narratives. Trawling was a man's world, yet beyond the dock walls, the fishing community was held together by a matriarchy. Thompson has noted the dependency of fishermen away at sea upon the work of women ashore.[1] In this narrative, the wives of trawlermen shouldered responsibility and enjoyed greater autonomy than was typical within the study period. But there were boundaries that women were not expected to cross. Hagmark-Cooper has observed the absence of women in maritime history. Where they do appear, it is often on exceptional

1. P. Thompson, 'Women in the Fishing: The Roots of Power between the Sexes', *Comparative Studies in Society and History*, 27:1 (1985).

or incidental terms.[2] A comprehensive and theorised historical analysis of gender roles and experience within British fishing is yet to be delivered. As a stepping stone, this work blends the voices of men and women into a holistic account of community, that bridges gendered perspectives as it reaches across land and sea.

In terms of demographics, the profile of narrators includes three born before 1930, one (interviewed regarding heritage) after 1970 and seven omitted to provide a date of birth. The remaining majority were born between 1931 and 1940 (12) and between 1941 and 1950 (15), placing them in the middle of their careers when crisis hit the fishery. Five were born between 1951 and 1960, making them the last generation to engage with a large-scale trawl industry in the city. Five narrators worked in trawling or fisheries management; 15 were trawlermen; five worked in the fish trades (sales and processing); and seven worked in ancillary trades. In addition, six narrators were interviewed as the female relatives of trawlermen (wife, sister, daughter); three entirely on the basis of their long-time residency on Hessle Road; one as a benefits officer; and five in relation to trawling heritage only. Twenty-two narrators had lived on Hessle Road at some point in their lives (16 male, six female).[3] Of the 15 trawlermen interviewed, 11 were one-time Hessle Roaders (for the six female relatives the figure was four). Seven narrators had first-hand experience of housing clearance, meaning that they or a close relative had been relocated due to demolitions occurring within the study period.

In this book narrators speak through a written representation of their spoken words. Transcription from oral sources presents a longstanding quandary for oral historians. Theorists reflect upon the loss of vitality, authenticity and meaning when sound is translated to written form. Good concedes that 'we must learn to live with the fact that transcription of the spoken word is more of an art than an exact science', and he advocates making the process as transparent as possible.[4] In my own transcription process, capturing local accent became important as the 'voice' in which

2. H. Hagmark-Cooper, 'To be a Sailor's Wife: Ideals and Images of the Twentieth-Century Seafarer's Wife in the Aland Islands', in B. Beaven, K. Bell and R. James (eds), *Port Towns and Urban Cultures: International Histories of the Waterfront, c.1700–2000* (London: Palgrave Macmillan, 2016), p.222.

3. The sample includes members of management who are more likely to live away from the area and heritage professionals unconnected with the district.

4. F. Good, 'Voice, Ear and Text: Words, Meaning and Transcription', in R. Perks and A. Thomson (eds), *The Oral History Reader*, 2nd edn (Oxon: Routledge, 2006), p.365.

most narratives had been expressed. Whilst avoiding the interpretation of vowel sounds, I have retained local words and speech patterns in a manner intended to respectfully capture the 'Hull voice'. In the final outcome, however, narrators speak through the author. Abrams has suggested that few oral historians feel entirely at ease with the academic manipulation of their respondent's words and I recognise that discomfort.[5] Oral history is an intimate approach to writing the past. In a spirit of co-production, my work aims to expose an 'original multiplicity of standpoints', which according to Thompson is the special contribution of the oral history method.[6] This is not a political, business or economic account, although it does involve all these perspectives. It is history co-created from human experience, through the assemblage of inquiry, interaction, subjectivity and memory. Ultimately however, it is the product of the historian, who takes control and responsibility for the representation.

Alongside memory, the work is underpinned by documentary research. With a limited historiography of post-Cod War British distant-water trawling, it soon became clear that primary archival research would have a substantial role in contextualising oral accounts. The archival sources consulted include industry management papers, technical reports, local authority records and trade journals. The key sources are from records of the Hull trawling firm Thomas Hamling & Company (operating between 1883 and 1983), held by the Hull City Archives (HCA). The documents consulted included the minutes, agendas and associated papers from the period 1975 to 1983 for the following organisations: the British Fishing Federation (BFF) (replacing the British Trawler Federation in 1976), a national organisation formed in 1918 to promote the interests of all aspects of the fishing industry; the Hull Fishing Vessel Owners Association (HFVOA), incorporated under the Companies Act in 1892 to represent owners' interests and to provide a range of common services; and the Hull Freezer Trawlers Owners Company (HFTOC), formed in 1979 prior to the liquidation of HFVOA and taking over some of that organisation's functions. Trade journals also offered critical insight into post-Cod War interests, events, developments and opinions from within the UK and international fishing industry. Journals consulted were: *Fishing News* (*FN*), 1975–82, *Fishing News International* (*FNI*), 1976–77 and *World Fishing* (*WF*), 1976–94.

5. L. Abrams, *Oral History Theory* (London: Routledge, 2010), p.165.

6. P. Thompson, *The Voice of the Past: Oral History*, 3rd edn (Oxford: Oxford University Press, 2000), p.6.

Beyond Trawlertown has a primary focus on the years after 1976, but the work spans a broader timeline. Chapter 2 visits Hull's fishery in the 1950s and 1960s, providing the baseline for understanding subsequent change. The emergence of the freezer fleet in the late 1960s is often overlooked in popular memory, but it is included here as a technology that was disrupting established patterns of trawling in advance of the crises that were to come. Chapter 3 introduces the challenges faced by the British distant-water industry in the 1970s, along with attitudes and expectations from within the Hull fishery. Chapter 4 examines the lived experience of contraction and decline in the immediate post-Cod War era, whilst Chapter 5 explores the opportunities pursued by Hull trawling firms and trawlermen in the decade up to 1985, and the new ways of living that these induced. For organisational purposes, the structure separates the experience of new beginnings from narratives of decline, but in practice the two occurred together. By 1985, the scale and character of Hull's fishery had been fundamentally revised, and the surviving industry entered a phase of operation that is largely outside the scope of this work. However, Chapters 6 and 7 offer interdisciplinary and cultural insights that take the narrative further. Chapter 6 lightly employs theories of lived place to examine the remembered reconfiguration of port city landscapes and relationships. Chapter 7 looks beyond 1985 and considers the transition from active industry to maritime heritage as an integral part of the story. In a 'history of heritage' it draws upon and contributes to critical heritage studies, revealing how the collective experience of rapid loss gave rise to a grassroots campaign to find the 'right place' for the trawling past in the evolving present-day city. Collectively these chapters do not seek to offer *the history* of British distant-water trawling in the post-Cod War era, but *a history*. It is a narrative channelled through the experience of a single trawling port, compiled from individual and collective memory, framed by documentary research and expressed through the understanding of the historian. For those whose lives it represents, it offers a version of their past that may be recognised or contested.

Chapter One

Introduction

In 1970, at the opening of a new decade, Michael's career at sea had already spanned 14 years. As a distant-water trawler skipper sailing from Hull on the North-East coast of England to the fishing grounds of Iceland, Michael was accustomed to facing challenging conditions to fill a ship with fish. He had an established pattern of living, with 21 days at sea followed by just three days at home. In the industry tradition, Michael had worked his way up from the deck to achieve his position. Ahead of him were further years in a job where hard work and danger were offset by the potential high earnings of the trawler skipper and the satisfaction of working in a familiar role. Michael was happy with his way of life and had no reason to think that it would end. As a deckhand, Thomas too enjoyed life at sea. In 1970, Thomas was just starting out in trawling, working up to 18 hours on the open decks of the old sidewinder fleet. He had ventured on the new freezer trawlers that were beginning to remould the distant-water industry. Thomas had toyed with a career in merchant shipping but had returned to trawling where he could work with his mates. Like Michael, Thomas looked ahead with little concern.

Ashore, Margaret lived and worked within the time-served parameters of an established fishery. Growing up in the dockside district of Hessle Road, trawling had influenced her life from birth. Her father worked on the fish dock as a 'bobber', swinging fish-laden baskets from waiting trawlers aligned along the quay. Her brother was at sea, fishing in the waters of the North-East Atlantic. Margaret herself worked at times in the nearby factories processing the catch landed by the city's own fleet, or she served in the local pubs frequented by trawler crews enjoying their brief time ashore. Margaret was happily immersed in a close-knit and at times fatalistic world that she did not imagine would change.

Into this atmosphere of permanence there came disturbance. At first, to Michael, Thomas and Margaret, caught up in the everyday flows of trawling, the pending threats perhaps seemed peripheral. However, the 1970s were to become a decade of contraction within many established industrial pillars of the British economy, with far-reaching social and cultural consequences. After the Second World War, Britain's post-war reconstruction boom had been followed by full employment and rising standards of living. In the 1960s, the favourable conditions were beginning to weaken in the face of global competition, balance of payments crises, wage inflation and a general decline in profits. These factors were compounded in the following decade by supply-side shocks, which included the soaring oil prices of 1973–74 and 1978–79.[1] In scholarly discourse, as well as in popular memory, the 1970s have been defined as economically troubled.[2]

Britain's maritime sector faced similar difficulties. A growing crisis in shipping, shipbuilding, fishing and ports emerged from as early as 1958, and from 1973 Britain's relative performance in the maritime sector was declining rapidly.[3] Conservatism and lethargy within shipping and shipbuilding have been named as culprits by both contemporary and later observers. Re-evaluation, however, gives a role to globalisation and growing international competition, which displaced British and Western European dominance in a maritime monopoly that could not last.[4] In shipping, the mid-1970s proved pivotal, with the oil crises accelerating an already downward trend in a maritime sector where oil costs were paramount.[5] In British ports too, the 1970s saw the ongoing erosion of the status quo.[6] From the 1960s, radical change was occurring with the advancement of containerisation and roll-on-roll-off transport, which merged with labour reforms that were revising the casual nature of dockside employment.

1. R. Coopey and N. Woodward, 'The British Economy in the 1970s: An Overview', in R. Coopey and N. Woodward (eds), *Britain in the 1970s: The Troubled Economy* (London: UCL Press, 1996), pp.4–6.
2. Cooper and Woodward, 'British Economy', p.1.
3. A. Jamieson, 'An Inevitable Decline? Britain's Shipping and Shipbuilding Industries since 1930', in D. J. Starkey and A. Jamieson (eds), *Exploiting the Sea: Aspects of Britain's Maritime Economy Since 1870* (Exeter: University of Exeter, 1998), pp.79–80.
4. S. Tenold, 'The Declining Role of Western Europe in Shipping and Shipbuilding, 1900–2000', in N. Petersson, S. Tenold and N. White (eds), *Shipping and Globalization in the Post-War Era* (Open Access: Palgrave Macmillan, 2019).
5. Tenold, 'Declining Role', p.29.
6. J. Byrne and A. Ombler, 'Memory on the Waterfront in Late Twentieth-Century Hull', in D. Starkey, D. Atkinson, B. McDonagh, S. McKeon and E. Salter (eds), *Hull: Culture, History, Place* (Liverpool: Liverpool University Press, 2017), pp.279–297.

Despite the benefits of modernisation, the process was accompanied by a substantial reduction in the workforce. Combined with the fallout of de-industrialisation and a diminishing maritime economy, it brought social, economic and spatial challenges to British port cities.

Against this backdrop of late-twentieth-century maritime disruption, Britain's distant-water trawl industry faced its own struggles. In 1976, Britain was defeated in a longstanding political and seaborne conflict with Iceland known as the Cod Wars. The dispute was about fishing rights. From 1952, the island had expanded its territorial waters to include an area that extended first four, then 12, then 50 and finally 200 miles from its coastline. As other nations followed, exclusion from habitual fishing grounds combined with rising oil prices, falling fish stocks and a new global politics of fishing to produce a perfect storm, which threatened the future of the British distant-water industry.

The significant investment demands of distant-water trawling had ensured its concentration in just a handful of British ports. Hull was one such locality.[7] Positioned at the confluence of the rivers Humber and Hull, towards the North Sea coast of England, the city had engaged in maritime trade, shipping and shipbuilding from the middle ages. From the mid-nineteenth century and in the wake of the city's declining whale trade, fishing had added to its maritime diversity. Sustained trawling from Hull began with the discovery of fish-rich waters in the North Sea. But by the end of the nineteenth century, pressure on fish stocks and the arrival of new technology had propelled trawlers from the Humber towards the fishing grounds of the North Atlantic and Arctic seas. The substantial distant-water trawl fishery that emerged therefore grew from an interaction with and dependency upon productive marine environments that lay off the shores of other nations. Almost a century later, the loss of these grounds was to trigger a series of events that would force British distant-water trawling into a rapid contraction and decline. As the impacts echoed between land and sea, the local consequences of global actions were made clear. International politics and globalised fishing brought change directly to Hull's door and the repercussions were felt by Michael, Thomas, Margaret and others like them, whose lives were determined by trawling.

7. The British distant-water industry was dominated by the Humber ports of Hull and Grimsby, along with the Lancashire port of Fleetwood. On a far smaller scale, distant-water vessels also operated at times from Lowestoft, Milford Haven, Swansea, Aberdeen, Leith and North Shields.

Beyond Trawlertown is a journey through British distant-water fishing in an era of disruption, transition and decline. As a case study of the trawling port of Hull, it blends insight from oral history interviews and archival research to explore the response to the challenges facing the sector in the 1970s. The emphasis on Hull may suggest a local history. However, Hull's position amongst the world's leading distant-water pioneers gives the story national and international significance. Weaving memory and lived experience with documentary sources, the book goes beyond the established narrative. Personal recollections offer an alternative perspective on the past. Acts of remembering reveal how global events and marine policy impacted upon the seafaring community that lived with the consequences. In the wake of an active fishery, remembrance still resonates in the search for the 'right place' for the maritime past within the contemporary city. The Cod Wars may be regarded by some as an end, but the years that followed brought new beginnings. Through the convergence of time, place and memory, this work creates a holistic narrative of interweaving stories that lay bare the intricacies of the human interaction with the marine environment and the aftermath when its threads are broken.

Time

Time is an attribute of any history, but it is not always linear. Histories, and particularly those created from memory narratives, are dialogues with the past that are influenced by the present. From the moment of creation they are open to alternative or supplanting temporal perspectives, ushered by a constantly unfolding future. From our present-day platform, we employ the diverse interactions of memory to reach into a medley of non-sequential happenings. In the jumbled timescape of the city, Crang and Travlou observe how traces of multiple pasts mingle with the modern urban fabric to produce what they describe as 'cracks in the surface of the present where time can be otherwise'.[8] Similar disruption is evident in other time-based encounters. In this book, time may seem to take a linear journey, yet beneath the apparent logic lies the to-and-fro 'messiness' of remembering, where narrators make sense of their younger selves and former lives, and where current-day reflections sit alongside once contemporary 1950s

8. M. Crang and P. Travlou, 'The City and Topologies of Memory', *Environment and Planning D: Society and Space*, 19 (2001), p.175.

sociology, which is now itself an historical account. In this narrative, everyday photography, footage, documents and objects are uprooted from the past and replanted as history or heritage; plays and artworks look backwards for inspiration to ease difficult forward transitions; and in the background stands a dockside landscape where the persistence of industrial ruins serves to disrupt illusions of progress and order.[9] From such diverse engagements with the past this work makes its own fissures in the present and through interdisciplinary chapters, visits and revisits the overlapping layers of everyday life in a late-twentieth-century fishery.

Within British maritime history, the sea fisheries represent an under-explored area of research – something that Wilcox attributes in part to the industry's relatively small size within the UK economy.[10] Existing studies are therefore comparatively limited in number and largely restricted to business, technical, environmental or political accounts. Within these parameters, however, the port of Hull has an advantage. British distant-water trawling ports were few, and after the Second World War Hull operated an almost entirely distant-water fleet. As one of Britain's leading ports within this sector, at a time when Britain itself was a world-leading distant-water fishing nation, Hull features prominently in references to the industry in local, national and international analyses. The small collection of resources can therefore prove surprisingly rich. Two contributions are crucial to this book. In 1955, urban sociologist G. W. Horobin delivered the first academic study to detail the operation and culture of trawling in Hull, defining the close-knit and spatially contingent interdependency between the fish dock and its neighbouring community.[11] Occurring in the midst of planning for housing clearances on Hessle Road, Horobin assessed the occupational implications of spatial reorganisation and the dispersal of the local population, urging planners to take heed of the special circumstances created by trawling.[12] Horobin did not anticipate the pending loss of the fishery, but his observation of the neighbourhood dynamic at the height of the post-war industry make an important contribution to understanding the impact of what was to come.

9. C. DeSilvey and T. Edensor, 'Reckoning with Ruins', *Progress in Human Geography*, 37:4 (2012), pp.466–470.

10. M. Wilcox, '"To Save the Industry from Complete Ruin": Crisis and Response in British Fishing 1945–1951', *Business History* (2019), p.353.

11. G. W. Horobin, 'Community and Occupation in the Hull Fishing Industry', *British Journal of Sociology*, 8 (1957).

12. Horobin, 'Community and Occupation', pp.351–354.

Jeremy Tunstall's subsequent study *The Fishermen* was influenced by the work of Horobin. First published in 1962, *The Fishermen* offers a detailed and broad-ranging contemporary account of the life of Hull distant-water trawlermen based upon extensive interviews and participant observation, including three trips to sea.[13] *The Fishermen* stands alongside Dennis et al.'s *Coal is our Life* in a genre of 1950s and early 1960s studies that documented, asserted and celebrated working-class community in the face of perceived changes to the social order.[14] Both Dennis and Tunstall examined industries where occupation and community were closely entwined, bridging the two sociological genres of work and community studies. But Tunstall's work brought the added aspect of a community spanning sea and shore, where the line between work and community was further obscured by the operation of what he defines as an 'extreme occupation'. Although his study is not heavily theorised, Tunstall offers an interpretation of the motives and social forces behind the practices and attitudes that he observed. At times, he betrays the bias of a middle-class academic engaging in a working-class world, a position that he himself acknowledges. His work has been described as 'perceptive and unsentimental',[15] yet on occasion there is a distance from his subject that obscures the cultural richness conveyed by other writers. Stephen Piper, for example, served as a deckhand on Hull trawlers in the 1960s and 1970s and combines literary skill with experience and cultural insight to convey a technical yet at times moving portrait of life at sea and at home.[16] Robb Robinson, a prominent fisheries historian who grew up on Hessle Road, contrasts Tunstall's 'unremittingly bleak' depictions with his own personal recollections of a vibrant and colourful community.[17] Nonetheless, Tunstall's unflinching lens provides a thorough and sincere portrait of Hull's fishery in the late 1950s that is crucial to understanding contemporary culture and ensuing change.

13. J. Tunstall, *The Fishermen: The Sociology of an Extreme Occupation*, 2nd edn (London: MacGibbon & Kee, 1969).
14. B. Dicks, *Heritage, Place and Community* (Cardiff: University of Wales Press, 2000), p.90; N. Dennis, L. Fernando, M. Henriques and C. Slaughter, *Coal is our Life: An Analysis of a Yorkshire Mining Community* (London: Eyre & Spottiswoode, 1956).
15. R. Robinson, 'Hook, Line and Sinker: Fishing History – Where Have We Been, Where Are We Now and Where Are We Going?', *Mariner's Mirror*, 97:1 (2011), p.171.
16. S. Piper, *The North Ships: The Life of a Trawlerman* (Newton Abbot: David & Charles, 1974).
17. Robinson, 'Hook, Line and Sinker', p.171.

The Fishermen took a critical yet measured stance towards the management and operation of the distant-water industry. In 1968, however, Tunstall authored a Fabian Society pamphlet that openly condemned the distant-water fishery as Britain's 'most antiquated industry'.[18] The sector's labour relations, marketing, organisational and technological innovation, as well as governmental fishing policy, all met with scathing criticism. Twenty-first-century historians have offered more nuanced assessments. Reid, while confirming the contentious labour relationships, offers an alternative stance by placing the issues raised by Tunstall within a wider framework of political instability, market change and rational economic response that was evident within fisheries across the North Atlantic.[19] Holm has suggested that Tunstall's outlook, alongside the Marxist approaches of later observers such as Thompson et al. in *Living the Fishing*, have led to a pessimistic depiction of capitalist models of fishing, sometimes set against an idealised or optimistic view of traditional self-ownership models.[20] The influence of potentially 'value-laden' accounts in understanding the British distant-water industry in the late twentieth century is a platform for debate. What is beyond dispute is the definitive contribution of Tunstall's contemporary sociological critique of the single trawling port of Hull to the understanding of British distant-water fishing in the late 1950s. Sixty years later, the insights offered by *The Fishermen* have become the starting point for revisiting that same port community in the aftermath of disruption.

Place

Alongside fish, the dynamics of place are central to this narrative. Through memory, Hull's fishing district is articulated as a network of meaningful lived spaces. Phenomenological geographers studying the connection between people and their environments have observed the role of everyday engagements, social contact and the habitual collective movement of

18. J. Tunstall, *Fish: An Antiquated Industry*, Fabian Tract 360 (London: Fabian Society, 1968).
19. C. Reid, 'Britain's Most Antiquated Industry: Mr Tunstall and the Fishing Industry', *International Journal of Maritime History*, 22:2 (2010).
20. P. Holm, 'The Modernisation of Fishing: The Scandinavian and the British Model', in R. Fischer, H. Hamre, P. Holm and J. Bruijn (eds), *The North Sea: Twelve Essays on Social History of Maritime Labour* (Stavanger: Stavanger Maritime Museum, 1992); P. Thompson, T. Wailey and T. Lummis, *Living the Fishing* (London: Routledge & Kegan Paul, 1983).

bodies within space in the formation of place relationships.[21] The positive experience of lived space has long been associated with feelings of belonging and attachment, expressed more recently in terms of quality of life, life satisfaction and wellbeing.[22] To Relph, a deep sense of 'insideness' occurs when place is subconsciously encountered and understood from the interior, rather than observed from the 'outside'.[23] 'Insideness' is central to profound human relationships with their surroundings. Relph's 'authentic sense of place' means 'belonging to *your* place both as an individual and as a member of a community and to know this without reflecting upon it'[24] (original emphasis). Intimate, subjective and everyday engagements make up the 'lifeworld' of a group or individual. Seamon uses the term 'place ballet' to refer to the repetitious movements and daily routines that embed a strong, insider's sense of place.[25] Through such ongoing encounters, places are never completed, but always in the state of becoming as new interactions occur.[26] Place, therefore, is not fixed but fluid, subject to fluctuating practice, meaning and emotions. It can be renegotiated in the face of change.

The cultural turn in geography from the 1980s redefined places or landscapes as representations or texts that emit cultural or political messages. Representational place is socially constructed – the product of the power systems that have led to its creation and that determine the way in which it is viewed. Landscape may therefore be considered as something that 'conveys and cements certain ideological narratives'.[27] Careful decoding is required to reveal the assumptions, values or manipulations that are embedded in our surroundings and serve to normalise existing structures of dominance or authority.[28] However, from the

21. For example, E. Relph, *Place and Placelessness* (London: Pion, 1976); D. Seamon, *A Geography of the Lifeworld* (London: Croom Helm, 1979); Y. Tuan, *Topophilia: A Study of Environmental Perception, Attitudes and Values* (New York: Columbia University Press).
22. L. Scannell and R. Gifford, 'The Experienced Psychological Benefits of Place Attachment', *Journal of Environmental Psychology*, 51 (2017), pp.256–257.
23. Relph, *Place*, p.55.
24. Relph, *Place*, p.65.
25. Seamon, *Geography of Lifeworld*, pp.54–59.
26. A. Pred. 'Place as Historically Contingent Process: Structuration and the Time-Geography of Becoming Places', *Annals of the Association of American Geographers*, 74:2 (1984), pp.279–280.
27. J. Wylie, *Landscape* (Abingdon: Routledge, 2007), pp.71–72.
28. Wylie, *Landscape*, pp.71–72.

1990s, a growing dissatisfaction with the notion of landscapes as fixed, legible and defined by the workings of underpinning systems led to a resurgent interest in place as lived.[29] Non-representational theory emerged to explore the world once again from a position of within, focusing on the ordinary, ongoing, unconscious, routine-yet-unpredictable practices of everyday life.[30] Non-representational theorists emphasise performance, which is punctuated by affect; the push of the world upon the body and the pre-conscious experience of sensation, sound, smell. It is through the unplanned immediacy of life in the moment that place is experienced and ascribed with meaning. Critics of non-representational theory have questioned its emphasis on a pre-conscious or pre-personal self, which they say obscures the differentiated experience of class, race or gender. The contribution of its less politicised stance remains debated. Lorimer's conciliatory term 'more-than-representational' is often preferred as one that is conversant with wider geographical concepts, but nevertheless offers to explore 'a series of infinite "ands", which add to the world rather than extract stable representations from it'.[31]

Comprehending place as a meaningful encounter is significant to this history, which explores not only a disrupted fishery, but also a disrupted lifeworld. Chapter 6 draws upon Ingold's phenomenological concept of the taskscape to convey the human and spatial interactions of Hull's fishing district as Trawlertown.[32] The term 'Trawlertown' conveys the distinct and recognisable layer of connected, collective and constantly unfolding activity and meaning produced by the relentless rhythms of fishing. The fishing taskscape has both intangible and material form and exists as separate from, but at the same time located within, the wider city. In oral narratives it emerges as habitual encounters, a hint of otherness and a strong affiliation to a neighbourhood geared towards an extreme occupation.

29. B. Anderson and P. Harrison, 'The Promise of Non-Representational Theory', in B. Anderson and P. Harrison (eds), *Taking-Place: Non-Representational Theories and Geography* (Surrey: Ashgate, 2010), p.6; M. Rose, 'Landscape and Labyrinths', *Geoforum*, 33 (2002), p.460.

30. N. Thrift, 'Dead Geographies – and How to Make them Live', *Environment and Planning D: Society and Space*, 18 (2000).

31. H. Lorimer, 'Cultural Geography: The Busyness of Being "More-Than-Representational"', *Progress in Human Geography*, 29 (2005); L. Cadman, 'Nonrepresentational Theory/Nonrepresentational Geography', in R. Kitchin and N. Thrift (eds), *International Encyclopaedia of Human Geography* (Elsevier Science, 2009), p.456.

32. T. Ingold, 'The Temporality of the Landscape', *World Archaeology*, 25:2 (1993).

In the 1970s, the factors that forced Britain's distant-water industry into decline brought Trawlertown to its close. A lengthy process of housing clearances and redevelopment was already reshaping the spaces and patterns of life in the streets closest to the fish dock. The new technology of the freezer trawler was beginning to alter not only practices at sea, but also those ashore. On top of gradual change came the rapid closures and dispersals caused by the demise of the fishery itself. This final disordering saw the end of the connections that had brought Trawlertown into being. Relph observed the trauma that can arise from the breaking of spatial bonds and he quotes Cox (writing in the 1960s) in suggesting 'that there are many people 'who never fully recover' from the loss of 'continuity of relationships with places' that results from urban renewal projects'.[33] In this context he views the resistance of residents to redevelopment, even when presented with better physical accommodation elsewhere, as evidence of the depth of such relationships.[34] In the late twentieth century, Hull's trawling community faced just such disruption to the connections, activity and spatial relationships that formed their lifeworld. An exploration of how these place-based affiliations were formed brings a deeper understanding of the impact when those interactions ceased to be.

Memory: Remembering Time and Place

Where bonds have been retained it is in the realms of memory. Memory is the mainstay of this narrative, providing method and subject for exploring the recent past. In an oral history approach, the intricacies of remembering are engaged as an historical resource, with human experience at the core. Portelli describes oral history as having a 'different credibility'.[35] Rather than distort the past with subjectivity, as early critics of the method believed, it reveals instead the motives, attitudes and the emotional or cultural contexts that underpin the remembered act. Although seemingly about the past, remembering is an act that is performed in the present. Perspectives may therefore be fluid, responding to the time and circumstances in which they are told, and part of an ongoing endeavour to make sense of a personal and

33. Relph, *Place*, p.65.
34. Relph, *Place*, p.65.
35. A. Portelli, 'What Makes Oral History Different', in Perks and Thomson, *Oral History Reader*, p.37.

collective history. To Portelli, 'memory is not a passive depository of facts but an active process of creation of meanings'.[36] The meanings and values assigned to the past can be conveyed by what is remembered and what is forgotten, and more subtly by what is revealed and what is omitted. Just as with the formation of lived place, meanings can be revised as circumstances change.

Memory stories are socially constructed and exchanged – told in accordance with cultural and ideological conventions, and associated with the binding of identity and belonging.[37] For oral historians this means that received communal narratives and discourses are already embedded within the memories of their sources. Simultaneously, the circulation of individual oral histories are themselves contributing to a collective consciousness. Memory therefore is a two-way flow between the private and the collective in an ongoing cultural circuit. In his classic theory on collective memory, Halbwachs asserts that human memory works *only* within a collective context – within social groups such as family or community.[38] Individual memory may exist, but it is entirely merged with social context and group memory, and recollection that does not conform to the collective is doomed to fade.[39] Not everyone accepts the absolutes of Halbwachs's stance. Green prefers to focus on a 'dynamic relationship between inherited culture and the individual mind',[40] highlighting the capacity for individuals to critique and contest the dominant memory discourse with their own personal histories. Nevertheless, the creation of history from oral sources *relies* upon collectivity. Oral historians piece together a cumulative picture from individual accounts, whilst the social exchange and representation of the past has an influence on each individual narrative. This is at the core of the oral history method; that society can be examined through a singular testament and history can be written through collective narratives of individual human experience.

Remembering in oral history interviews can assume the role of bearing witness to the past as a means of handing it on. In James's *Dona Maria's*

36. Portelli, 'Oral History', p.37.

37. Abrams, *Oral History Theory*, p.109; D. James, *Dona Maria's Story: Life History, Memory and Political Identity* (London: Duke University Press, 2000), pp.124–125.

38. M. Halbwachs, *On Collective Memory* (London: University of Chicago Press, 1992), pp.172–173.

39. Halbwachs, *On Collective Memory*, pp.172–173.

40. A. Green, 'Individual Remembering and "Collective Memory": Theoretical Presuppositions and Contemporary Debates', *Oral History*, 32:2 (2004), p.41.

Story, Maria is aware that when her factory is pulled down there will be no record other than her testament of the struggles embodied there.[41] Within Hull's trawling community some narrators were similarly propelled by a sense that their past was forgotten and the desire for that to be remedied. Here, remembering emulates heritage as a means of attributing value to the past. It gives rise to the matter of authenticity. James observes a pact at the heart of the oral history interview. Due to the complexities of memory, narrators may not always be factually precise. But they must commit to *authenticity*; to recounting a personal truth that is based on 'fidelity to meaning'.[42] Authenticity matters because through oral history testament the past is invested with a significance that is bequeathed to the future. The relatedness of memory and heritage is often taken for granted in academic writing, but it is more rarely defined. Van Huis et al. present memory as an 'elastic concept' that is frequently extended to how memory and meaning are interpreted and expressed within socio-cultural practices and materialised forms.[43] Smith has presented heritage in similar terms, as a cultural process and something essentially intangible.[44] Heritage is not an object or something to possess, but an active and ongoing assignment of meaning to the past. Emotion, feelings and affect are embedded in the act. Wetherell, Smith and Campbell refer to heritage as an affective practice; one where emotion is a principal focus, and where the performance may uphold or disrupt 'comfortable' expressions of identity and belonging, with implications for the present.[45] From this perspective, heritage, memory, emotion, cultural knowledge, values, meaning and experience are entwined.[46] The vocabularies become inseparable.

Remembering, meaning and authenticity are integral to a portrait of Hull in the post-Cod War era. In the immediate aftermath of crisis, civic attempts to reinvigorate Hull's flagging economy initially overlooked the

41. James, *Dona Maria's Story*, p.153.
42. James, *Dona Maria's Story*, p.136.
43. I. van Huis, S. Kaasik-Krogerus, T. Lahdesmaki and L. Ellena, 'Introduction: Europe, Heritage and Memory – Dissonant Encounters and Explorations', in T. Lahdesmaki, L. Passerini, S. Kaasik-Krogerus and I. van Huis (eds), *Dissonant Heritages and Memories in Contemporary Europe* (Palgrave Macmillan, 2019), p.7.
44. L. Smith, *Uses of Heritage* (London: Routledge, 2006), pp.56–58.
45. M. Wetherell, L. Smith and G. Campbell, 'Introduction: Affective Heritage Practices', in L. Smith, M. Wetherell and G. Campbell (eds), *Emotion, Affective Practices and the Past in the Present* (Abingdon: Routledge, 2018), p.16.
46. Smith, *Uses of Heritage*, p.56.

traditions and spaces of Trawlertown. In the face of acts of forgetting, with buildings demolished and neighbours dispersed, the former fishing community began their own campaign for remembrance. Their story is a compelling example of what Robertson refers to as 'heritage from below'.[47] 'Heritage from below' is that constructed for, about and by local communities, with minimal professional intervention. The concept sits within an aspiration for a more democratised version of the past that gives primacy to a community's own values, sense of place and identity. In his seminal work *Theatres of Memory*, historian Raphael Samuel celebrates the eruption of what he terms 'unofficial knowledge', which counters academic and professional elitism.[48] From collecting vintage clothing to school-yard re-enactments, everyday activities are celebrated as valid ways to connect with the past. In the UK, Samuel equates the spread and acceptance of broader and more democratic concepts of heritage with the growth of grassroots movements from the 1960s, amplified in the 1970s by 'separation anxieties' created by 'the run-down of regional economies [*and*] by threats to the living environment which put the taken-for-granted at risk'.[49] The outcome, he says, was a new version of the national past; one consisting of 'little platoons' rather than the 'great society', sprung from 'a thousand hands' and 'inconceivably more democratic than earlier ones, offering more points of access to "ordinary people", and a wider form of belonging'.[50]

The view that everyone is entitled to engage actively with the heritage of their choice is increasingly recognised within governmental conventions and conservation principles.[51] However, putting such principles into practice can be problematic. The emergence of multiple voices claiming diverse pasts gives credence to an understanding of heritage as an essentially dissonant process. Tunbridge and Ashworth have presented the past as a resource in conflict.[52] The past, they say, is called upon to serve political, economic and cultural functions. Where such plural aims draw upon a single resource, the result can be dispute. Diversity of race, religion or

47. I. Robertson (ed.), *Heritage from Below* (Farnham: Ashgate, 2012).

48. R. Samuel, *Theatres of Memory, Vol. I* (London: Verso, 1994).

49. Samuel, *Theatres of Memory*, pp.145–150.

50. Samuel, *Theatres of Memory*, p.158, p.18 and p.160.

51. J. Schofield, 'Heritage Expertise and the Everyday: Citizens and Authority in the Twenty-first Century', in J. Schofield (ed.), *Who Needs Experts: Counter-Mapping Cultural Heritage* (Farnham: Ashgate, 2014), p.5.

52. J. E. Tunbridge and G. J. Ashworth, *Dissonant Heritage: The Management of the Past as a Resource in Conflict* (Chichester: Wiley, 1996).

class adds further disparity, leading Tunbridge and Ashworth to construe heritage as intrinsically dissonant.[53] Smith goes further, seeing dissonance as the very essence of heritage, the defining characteristic of the unending process of assigning meanings to the past.[54] In the negotiation, some are excluded in acts of disinheritance.[55]

Democratisation and dissonance are components within Robertson's concept of 'heritage from below'. Robertson introduced the term to dislodge the dominant academic emphasis on processes of heritage that respond to national narratives and identities. In cultural terms, 'heritage from below' is a counter-hegemonic process. It can take the guise of an alternative vernacular discourse: the attempts by communities to define and engage with their own histories.[56] These are Samuel's 'little platoons', born of enthusiasm, curiosity or belonging. Such expressions gently subvert the dominant discourse, not in open conflict, but with the suggestion of other possibilities. Elsewhere, however, 'heritage from below' may be the cause or tool of an active challenge and push for change. As community protest, it underpins the rise of fishing heritage groups in Hull, who have contested the perceived erasure of their past.

'Heritage from below' as social action returns us to the intricacies of dissonance. Heritage dissonance is most commonly presented as occurring between powerful elites, that assert partial pasts for political gain or expert curation, and the dominated groups that challenge them. Yet its scope extends further. Dicks has shown how the heritage value of local spaces can be contested within communities, with a source of pride for one existing as a site of loss or a backward-looking eyesore for another.[57] Robertson's own study on the Hebridean island of Lewis reveals how a community-led project to mark the sites of local historical land protests ran into strong internal disputes between settlements, resulting in breakaway groups and the erection of alternative monuments. Even at this most local level, where the past is under negotiation, contestation is part of the process.[58]

53. Tunbridge and Ashworth, *Dissonant Heritage*, p.21.
54. Smith, *Uses of Heritage*, p.82.
55. Tunbridge and Ashworth, *Dissonant Heritage*, p. xi.
56. Robertson, *Heritage from Below*, pp.7–8.
57. Dicks, *Heritage, Place*, pp.160–169.
58. I. Robertson, 'Heritage from Below: Class, Social Protest and Resistance', in B. Graham and P. Howard (eds), *The Ashgate Research Companion to Heritage and Identity* (Aldershot: Ashgate, 2008), pp.150–156.

The implications are significant and often overlooked. With its inherent dissonance and struggles for control, 'heritage from below' shares characteristics with that imposed from above. Muzaini and Minca have cautioned against approaching 'heritage from below' as a value-free process, pointing out its own susceptibility to ulterior motives, inherent bias and partial narratives.[59] Once memory seeks representation in the cultural arena, it must stake its claim robustly. Crooke suggests that in the community museum, by producing an exhibit, communities take control of their own narratives and attempt to manage how others see them. Just as with expert or elite representations, this is not a value-free process, but one that is selective. The representation draws an audience into the community message, disseminating it further.[60] In this sense, community representations, democratised and 'from below', become endowed with a certain power, and this raises issues and consequences for those whose role it may be to mediate.

Concepts of place, memory, dissonance and 'heritage from below' have all been part of the transition of Hull's distant-water trawl fishery from an active industry to maritime heritage. Understanding the culture and practices of the operating fishery and the impact of rapid change precedes a deeper appreciation of how and why fishing heritage continues to be important in the twenty-first century. Heritage provides a still unfolding 'last stop' in the journey through Hull as a leading British distant-water trawling port in a climate of late twentieth-century decline and transition. It is a journey evoked by personal and communal recollections of an extreme maritime occupation. Lightly framed by interdisciplinary perspectives, this book portrays a changing relationship with the sea. In its entirety the portrait is unique, assigned to this time and this place. In part, it offers insights that transfer to other port cities, marine encounters and working-class districts. At its heart is the divorce of a maritime community from the industry that gave it existence. It is underpinned by an affective and compelling desire to remember.

59. H. Muzaini and C. Minca, 'Rethinking Heritage, but "From Below"', in H. Muzaini and C. Minca (eds), *After Heritage: Critical Perspectives on Heritage from Below* (Cheltenham: Edward Elgar Publishing, 2018), p.12.

60. E. Crooke, 'The Politics of Community Heritage: Motivations, Authority and Control', *International Journal of Heritage Studies*, 16:1–2 (2010), pp.27–28.

Chapter Two

Trawling Lives and Culture – A Portrait up to 1976

The Trawling Port of Hull

In Hull's long history as a port city, the episode of trawling is relatively short. From inception in the mid-nineteenth century, to crisis and decline in the mid-1970s, the period spanned less than 150 years. Within that time the city became a pioneer of distant-water trawling in the North Atlantic. On the eve of disruption Hull stood amongst the leading distant-water fishing ports of the world. The city's sustained involvement with trawling emerged from the 1830s, when Devon fishermen increasingly made seasonal visits to the north-east coast. Their activity led to the opening up of the rich 'Silver Pits' fishing grounds to the east of the mouth of the Humber Estuary. From the mid-1840s southern smack owners took full advantage of the new discovery, making a temporary base in the Humber ports of Hull and Grimsby. As the incomers began to settle permanently, names such as Richard Hamling, Robert Hellyer and Thomas Boyd established owner dynasties that would become synonymous with the Hull fishery.[1] Rail transport was essential to the expansion of Hull's fishing industry. From 1840, the Hull and Selby Railway eased access between Hull and the industrial centres of the North of England. As the rail network expanded, the mass markets of London and the Midlands became accessible to the Humber ports.[2] Low-cost transportation linked the coastal supply of fish with the inland need for cheap nourishment for the poor working-class populations of rapidly expanding industrial

1. R. Robinson, 'The Line and Trawl Fisheries in the Age of Sail', in D. J. Starkey, C. Reid and N. Ashcroft (eds), *England's Sea Fisheries: The Commercial Sea Fisheries of England and Wales since 1300* (London: Chatham, 2000), p.78.
2. R. Robinson, *Trawling: The Rise and Fall of the British Trawl Fishery* (Exeter: University of Exeter Press, 1996), pp.26–28.

centres. From the mid-nineteenth century, the growing popularity of fried fish further stimulated demand and by the turn of the twentieth century, fish and chips were well established in the national diet.[3] For Hull, the growing market for cheap fish and chips stimulated a swift 'demand-driven transformation'.[4] By 1863, Hull had 270 sail fishing smacks, increasing to 420 by 1880.[5] From the 1870s vessels fished further from the British coast under a 'boxing' system, where a fleet remained at sea for extended periods, with a fast cutter running ashore daily with boxes of fresh fish.

From the 1880s, the adoption of steam trawling in Hull cemented the city's future as a distant-water fishing port. The transition from sail to steam was a business response to evidence of declining fish stocks in the North Sea.[6] The triple expansion engine, accompanied by the local manufacture of ice gave Hull vessels the range and capacity to venture to distant and less exploited grounds. The steam trawler needed capital investment beyond the reach of most smack owners, and the way forward was the limited liability company that could mobilise capital from outside the industry.[7] The engineering and financial demands of steam trawling ensured its concentration at larger ports like Hull, where capital investment, technological dependency and an established trawling ethos combined to produce a fishery that was henceforth 'quintessentially industrial in character'.[8]

By 1903, Hull's entire fleet was steam powered.[9] Although a North Sea steam boxing fleet operated until 1936, larger vessels increasingly fished singly in distant waters, beginning around Faroe and Iceland, then in the Barents Sea and followed later by seas at Spitsbergen and Bear Island.[10]

Before the First World War, trawlers had occasionally voyaged as far as Greenland and the Grand Banks of Newfoundland. Iceland, however,

3. J. Walton, *Fish & Chips and the British Working Class, 1870–1940* (Leicester: Leicester University Press, 1992).

4. Robinson, 'Line and Trawl Fisheries', p.80.

5. Tunstall, *The Fishermen*, p.18.

6. R. Mumby-Croft and M. Barnard, 'An Antiquated Relationship? Trawler Owners and Trawlermen, 1880–1980', in Starkey et al., *England's Sea Fisheries*, p.120.

7. R. Robinson, 'Steam Power and Distant-Water Trawling', in Starkey et al., *England's Sea Fisheries*, p.210.

8. Robinson, 'Steam Power', pp.209–210.

9. M. Thompson, *A Tribute to Hull's Fishing Industry* (Beverley: Hutton Press, 1996).

10. Robinson, 'Steam Power', pp.211–212.

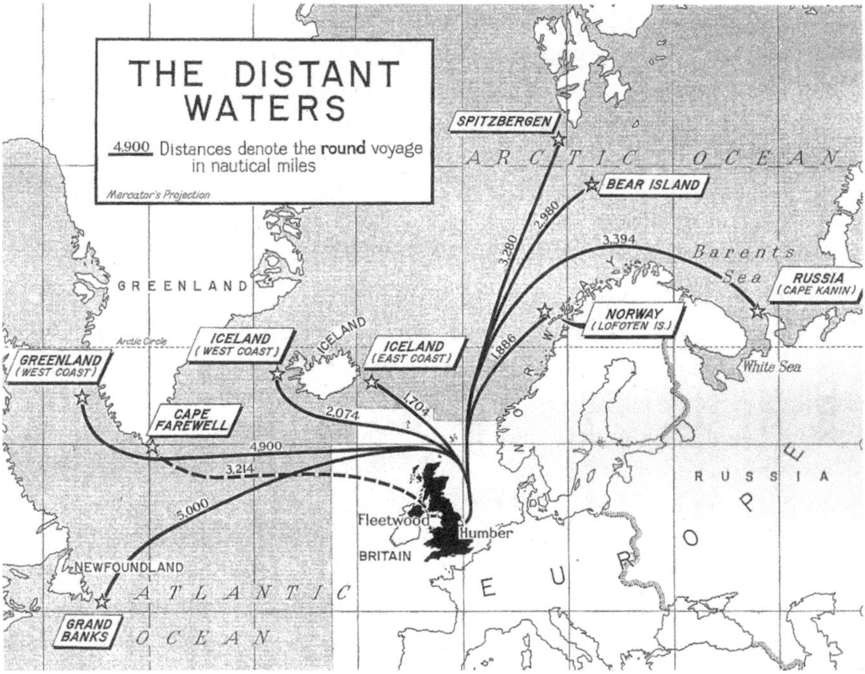

Map 1. The distant-water fishing grounds.

remained a favourite ground and as early as 1903, 80 Hull trawlers were fishing in Icelandic waters, establishing the 21-day trip that defined the operation and culture of the Hull industry for the next 60 years.[11] The first three decades of the twentieth century further cemented the connection between Hull, cod and the fish and chip trade. Fish taken by Hull trawlers at Iceland was coarser than that of the North Sea, but used readily by fish friers who had less need for premium cuts. As the number of fried fish retailers grew throughout the 1920s and 1930s, Hull became established as the 'friers' port', dominating the market for the supply of lower quality white fish, mainly cod.[12] Throughout the interwar years Hull's fishery continued to progress, even as elsewhere in the UK the industry struggled with low returns.[13] From 1936, Hull's focus was almost entirely on distant

11. Robinson, 'Steam Power', p.211.
12. Walton, *Fish & Chips*, pp.44–45; Robinson, 'Steam Power', p.211.
13. Wilcox, 'To Save the Industry', pp.357–358.

waters, landing 45 per cent of England's white fish in 1938.[14] Yet as Britain headed towards the disruption of the Second World War, the catch from rich, distant Arctic grounds was creating oversupply, and from 1932, for the first time, the industry faced regulation and restriction.[15]

From the mid-nineteenth century, as Hull's new seafaring sector grew, so too did the networks, trades and facilities that comprised the wider fishery. Trawling summoned a sizeable population of workers and families with close ties to the diverse elements of the trade. Naturally, these sought to make their home close to the comings and goings of the fleet. In 1883, the fishing fleet moved into their permanent home at St Andrew's Dock, located to the west of the city, parallel with the Humber bank (Map 2).[16]

The fish dock was extended in 1897, and from these early years began to acquire the operational structure that changed little to its final days.[17] Maps 4 and 5 from 1951 reveal the enduring character of the dock that was established 60 years prior. Fish were landed into open-fronted sheds on its northern side. Known as the wet side, this also served as the fish market, where fish would be weighed into containers known as kits and sold using the method of the Dutch auction. From 1899, cod was also dried and salted at the 'cod farm', situated west of the wet fish market.[18] At the western end of the dock was the slipway for trawler surveys and repairs, and behind the market sheds an entanglement of tracks enabled the North Eastern Railway to transport fish from the quay.[19]

14. Robinson, 'Steam Power', p.213.

15. Robinson, 'Steam Power', p.213; Wilcox, 'To Save the Industry', pp.357–358.

16. Map 2 shows the study area in the nineteenth century, relative to the rest of the city. St Andrew's Fish Dock and the streets of Hessle Road lie to the west, parallel with the River Humber. The city's central docks form a ring around the Old Town, and to the east lie the commercial port and adjacent neighbourhoods. Map 3 shows the physical relationship of dock and district. In the proximity of the fish dock are the densely packed courts and terraces that were home to fishermen, dockworkers and fish processors.

17. M. Thompson, *Fish Dock: The Story of St. Andrew's Dock Hull* (Beverley: Hutton Press, 1989), pp.13–14.

18. Thompson, *Fish Dock*, p.15.

19. Map 5 the close-up gives detail of the entrance to St Andrew's Dock. The fish market (Billingsgate) is to the north, the larger ice factory to the south and around the lock gates is a cluster of offices, stores and works. The distinctive Lord Line building, which survives at the eastern end of the dock, was opened in 1949 and is shown here as a dotted outline.

Map 2. 1893. Kingston upon Hull, Hessle Road and St Andrew's Fish Dock.

Source: Landmark Information Group, *Kingston upon Hull, Yorkshire*. 1:10560 (1st revision). Ordnance Survey County Series, EDINA Historic Digimap Service, 1893, http://digimap.edina.ac.uk (created 26 September 2015).

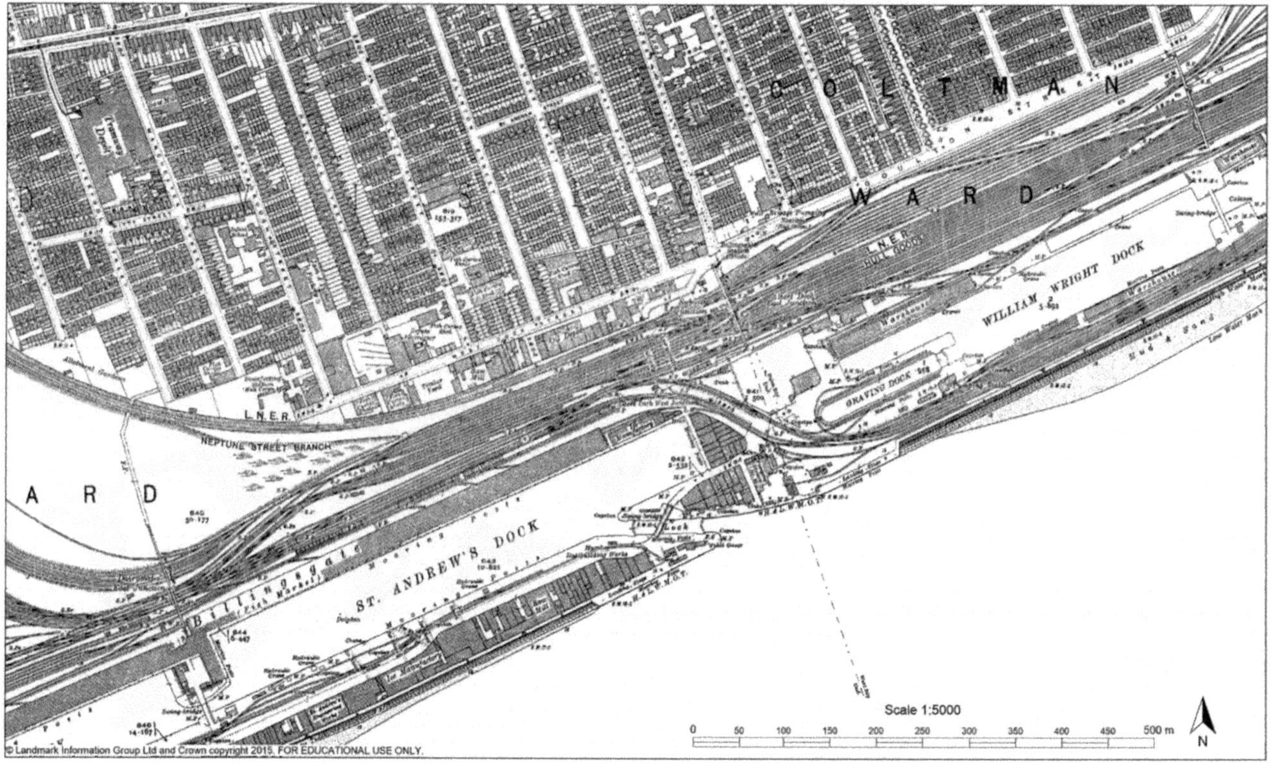

Map 3. 1928. Hessle Road and St Andrew's Fish Dock.

Source: Landmark Information Group, *Kingston upon Hull, Yorkshire.* 1:2500 (2nd revision). Ordnance Survey County Series, EDINA Historic Digimap Service, 1928, http://digimap.edina.ac.uk (created 26 September 2015).

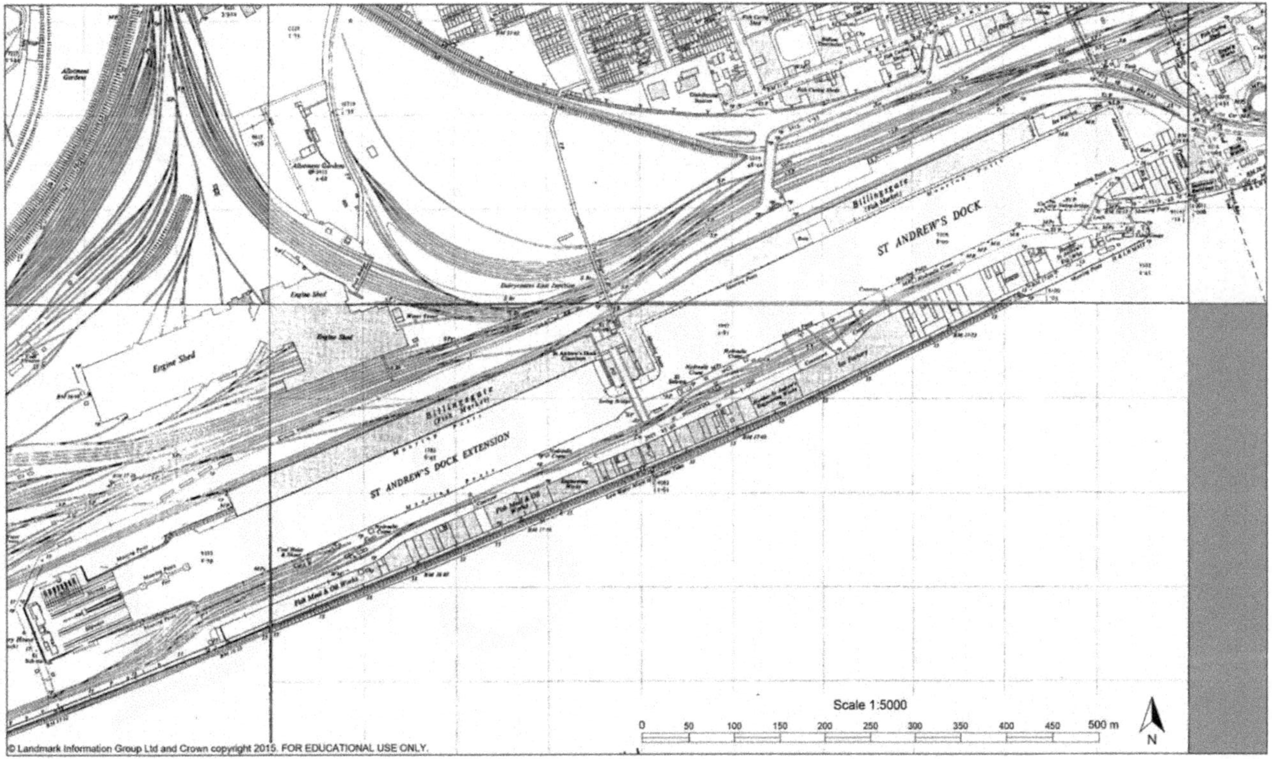

Map 4. 1951. St Andrew's Fish Dock.
Source: Landmark Information Group, *Kingston upon Hull, Yorkshire*. 1:2500. National Grid Tile TA0727, EDINA Historic Digimap Service, 1951, http://digimap.edina.ac.uk (created 26 September 2015).

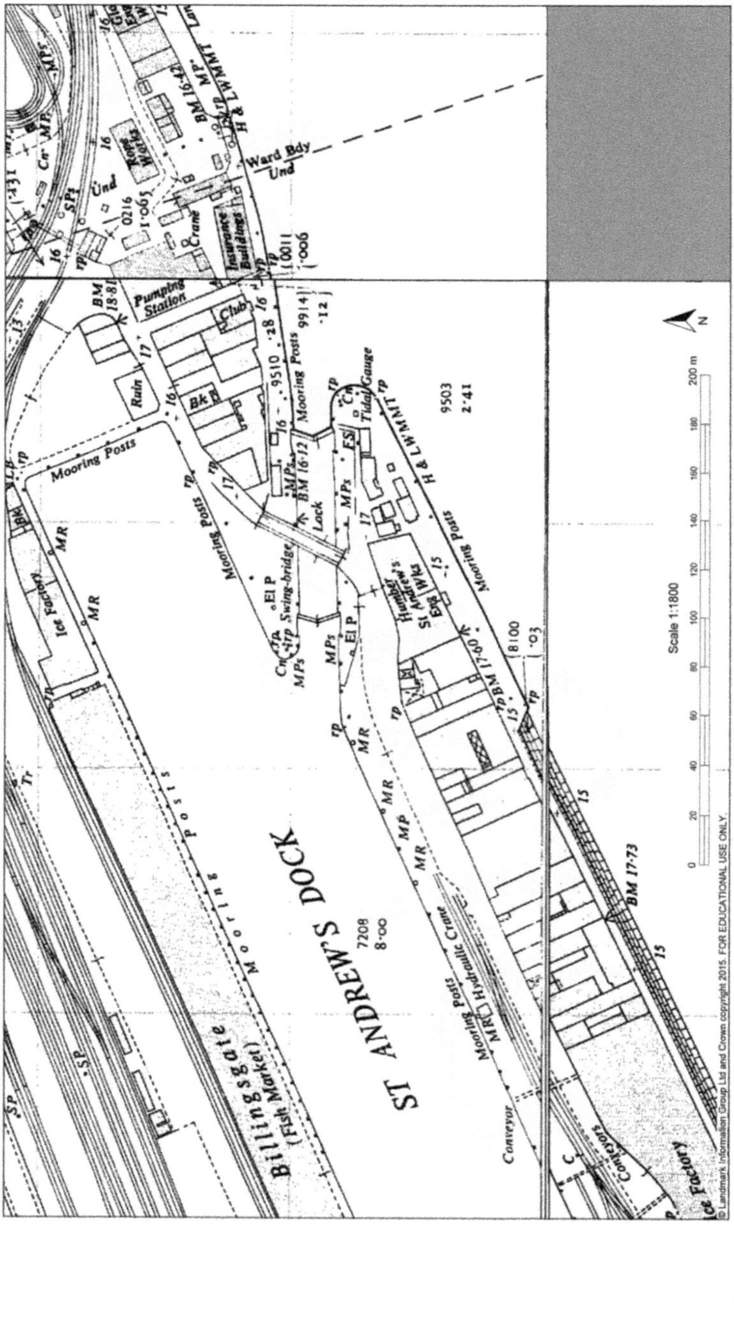

Map 5. 1951. Close-up of the lockpit area, St Andrew's Fish Dock.
Source: Landmark Information Group, *Kingston upon Hull, Yorkshire*. 1:2500. National Grid Tile TA0727, EDINA Historic Digimap Service, 1951, http://digimap.edina.ac.uk (created 26 September 2015).

To the south was the dry side, home to engineering and repair sheds, equipment stores and the premises of trawler firms and other related companies. Offices, stores and cafés were further concentrated around the eastern limit of the dock, surrounding the lock gates that led out into the Humber. In 1890, the Hull Fish Meal and Oil Company built its fish offal factory, followed in 1891 by the Hull Ice Company plant.[20] Along this 'dry side', ice was taken on board, crews signed on for trips and ships were serviced, fuelled and supplied ready to go to sea. By the 1930s, the Hull Fishing Vessel Owners Association (HFVOA) had established its collective ownership of the ice factories, fish meal plant and other important secondary services.[21] The concentration of facilities, as well as vessels, into the hands of the HFVOA ensured that trawler owners had considerable interest in the success of Hull's fish dock. Their successful common-sense business approach meant that trawling in Hull developed as a highly localised industry, instilling a strong disincentive to operate away from the port.

Beyond the tracks of the North Eastern Railway, a different expansion kept pace with the burgeoning docklands. North of the fish dock, the linear urban district of Hessle Road sprang up in response to increasing maritime trade and to trawling in particular. The semi-rural suburbs of the mid-nineteenth century were soon overtaken by the working-class housing that were homes for trawlermen and ancillary dockworkers. Map 3, from 1928, illustrates the density of life crammed into the courts and terraces of Hull's now established fishing district. Around the axle of Hessle Road an abundance of pubs, clubs, cinemas, shops and services provided for the fishing community, and in the streets to the south, residents lived at close quarters with the fish houses and processing facilities that spilled out from the dock.

There is a sense of carnival in descriptions of Hull's fishing district, even in its formative years. In 1883, the *Hull Critic* records Hessle Road as a 'vast, populous and growing locality'. The writer suggests that those wishing to travel through the area quickly should avoid the main thoroughfare, particularly on Saturday nights 'when the moving mass of people may be described as like unto a fair in numbers'.[22] Historian Robb Robinson evokes an atmosphere that he compares to a western gold-rush town, 'teeming with life and light'.[23] It was a far cry from the older fishing communities of the Yorkshire coast,

20. Thompson, *Fish Dock* p.14.
21. Tunstall, *The Fishermen*, p.36.
22. *Hull Critic*, 9 June 1883, quoted in E. Gillett and K. MacMahon, *A History of Hull* (Hull: Hull University Press, 1980), p.309.
23. Robinson, *Trawling*, p.129.

Figure 1. Easter fish landing, St Andrew's Dock, 1959. In a longstanding tradition, fish is packed into baskets fastened to ropes that drop down into the hold. The baskets are then swung ashore to be emptied into aluminium 'kits' waiting on the quayside (Copyright Innes.co.uk).

sharing much instead with the industrial working classes of the mill or mine. Yet as Robinson points out, it was also different. Although an industrial community in appearance, at its heart was the sea and the uncertainty of fishing, which evoked a sense of separation and cultural otherness.[24] As Gillett and MacMahon observe, with their homes concentrated at one end of the city, and with the exceptional working circumstances of trawling, 'for a whole century the fishing community was almost completely severed socially and geographically from the rest of Hull'.[25]

In 1939, Britain had entered the Second World War with its fisheries facing the problems of over-production and depressed prices. In the years immediately after, distant-water trawling enjoyed a transient boom. The war

24. Robinson, *Trawling*, p.130.
25. Gillett and MacMahon, *History of Hull*, p.309.

had given fishing grounds chance to recover from over-exploitation, and post-war rationing and price controls ensured a ready demand for fish.[26] By 1949, however, ageing vessels, low fish prices and the end of food shortages were creating a crisis for the UK near- and middle-water sectors, which from 1951 was alleviated by government subsidy, and with regulation, reorganisation and modernisation overseen by the newly formed White Fish Authority (WFA).[27] Despite also facing difficulty, due to their concentration, coherence and potential to self-regulate the distant-water sector was able to forge ahead and 'pay their own way'.[28] As a result, they were excluded from government subsidies for modernisation that were offered to near- and middle-water vessel owners. Instead, the distant-water fleet modernised without government assistance. In Hull, concentration of ownership into 15 to 20 large trawling firms helped to secure the resources required for new vessels and new technologies.[29] From 1946, coal-fired trawlers gave way to oil and from the mid-1950s diesel and diesel-electric began to take over from steam.[30] Modern electronics such as radar and echo sounders added to the efficiencies brought by larger vessels and powerful engines.[31] As new technology was embraced, older technologies were gradually phased out. The process meant that at any point in time, the Hull distant-water fleet included trawlers of varying age and advancement. Investing in new vessels brought risk and financial cost, and the pace of change was steady and considered.

A Trawling Life

The Practice and Culture of the 21-day Trip

> Three weeks at sea, three days at home.
> Victor, trawler skipper

Until economics, technology and eventually politics forced new ways of operating, life in Hull's fishery followed long-established rhythms.

26. Robinson, *Trawling*, p.186.
27. Wilcox, 'To Save the Industry', pp.359–377.
28. Robinson, *Trawling*, pp.187–188.
29. Robinson, *Trawling*, p.189.
30. Robinson, 'Steam Power', p.214.
31. J. Thor, *British Trawlers and Iceland, 1919–1976* (Goteborg: University of Goteborg, 1995), p.114.

Describing work at sea in 1960, Tunstall observed that group life on a trawler had an 'absolute and unchanging quality'. With only minor variations, it is, he says, 'the same as the old fishermen remember', and he suggests that this 'gives the fisherman a certain feeling that whatsoever exists in the life on a trawler is somehow permanent and ordained'.[32] In Hull, as in other British distant-water ports, the practices of trawling had grown up around the 21-day trip. Twenty-one days was the voyage limit of the traditional sidewinder trawler; the time needed to journey to the North-East Atlantic, fish for around ten days and return with the catch still fresh enough for market. The pattern emerged from the early days of distant-water trawling. It is recorded in the interwar diaries of Hull skipper William Oliver, was observed by Tunstall in the 1950s and 1960s and continued within the substantial fleet of side trawlers still operating into the 1970s.[33]

The dominance of the 21-day trip influenced not only life at sea, but also that of the dockside and surrounding district. The profit demands of trawling meant that ships spent minimal time in port. In the early 1950s, trawlers docked for just 48 to 60 hours. By the late 1960s, the turnaround time of the sidewinder fleet had increased only marginally to 60 to 72 hours.[34] Life for the trawlerman at home was compressed into three days and consequently assumed a frenzied and celebratory nature. Vessels always returned to Hull and the regular, yet frenetic, comings and goings forged a distinct culture and tradition within the portside community. When Alan H. moved to Hull in the 1960s to work for the WFA, he was struck by the contrast between trawling and his experience of the Merchant Navy. He observed a tough life that required a certain character and expectation:

> the families had to accept that the man of the household would be away for three weeks and life had to carry on when he was not there. And when he came home, life would change for about three days and then maybe peace would reign for another three weeks until he came

32. Tunstall, *The Fishermen*, pp.133–134. Tunstall's observations were collected during three trips aboard Hull trawlers, one taking place in 1958 and two in 1960, see p.280.
33. D. J. Starkey, 'Distant-Water Trawlerman: William Oliver, 1884–1959', in D. J. Starkey, D. Atkinson, B. McDonagh, S. McKeon and E. Slater (eds), *Hull: Culture, History, Place* (Liverpool: Liverpool University Press, 2017).
34. Robinson, *Trawling*, p.191.

Figure 2. *D. B. Finn* (Hull) and *Arkinholm* (Fleetwood) – side trawlers, known as 'sidewinders', moored in front of the Lord Line trawler offices, St Andrew's Dock, 1974. (Copyright Alec Gill).

> back again. So, it was a lifestyle that I think a lot of other communities could not adapt to. It was something you lived with, grew up with and accepted.[35]

Fitting home life into three days demanded energy. Victor, a trawler skipper, describes life at home as hectic, recalling how 'a lot of stuff' had to be crammed into just three days.[36] Michael outlines a typical spell ashore in the late 1960s:

> Well in three days you had to cheer the family up, and the little 'uns and like spoil them for a bit cos they don't see their father for three weeks ... you'd take your wife shopping up town and spoil her for a couple of hours and sit in shops fed up. And then the kids'd come home from school and you'd have bought them some presents or you'd take

35. Interview with Alan H, November 2012.
36. Interview with Victor and Pauline, July 2013.

> them up town … and the next day you relaxed a little bit and then the day after that you was away to sea again …[37]

Socialising and leisure were important during time ashore. Local memory is rife with tales of epic celebration, drinking and spending, as trawlermen, short on time and flush with trip-money, earned themselves the reputation of the 'three-day millionaires'.[38] However, the notion of the 'drunken trawlerman', must be viewed within the context of the 21-day trip. To Jim, a trawler skipper, drink-fuelled escapades ashore were a natural response to the confines of life at sea:

> If you ask anyone ashore – stop in house, don't go out, sleep, work 18 hours a day, do it for three weeks and at the end of the three weeks … the boss says, there you are, there's your three weeks money, but you're back on Tuesday … what you gonna do with it [laughs]?[39]

Whilst heavy drinking caused problems within some households,[40] for others, time ashore was for family and going out with wives and friends. Socialising took place in many parts of the city and beyond, but activity often focused on Hessle Road; the heart of the fishing community, where the daily arrival of ships and crews gave the area a continual and collective buzz.

The impetus for intense and jubilant shore time is better understood when set against the experience of life at sea. Arctic fishing was demanding and dangerous. It was a physical job, employing heavy gear on a moving deck, in freezing and hostile seas. The risk of accident was high, and hazards were compounded by the operational practices and culture that were deeply embedded within the side trawler fleet. The typical sidewinder crew numbered 20. Whilst fishing, men worked on an open deck, usually for ten days and on a rota of 18 hours' work to six hours' sleep. Tunstall believed that dampening cycles of fatigue focused attention on immediate

37. Interview with Michael, November 2012.
38. A. Gill, *Three-Day Millionaires: Trawlermen Home Between Trips* (Hull: Dovedale Studios, 2001) (DVD). The term refers to the tendency for distant-water trawlermen to spend earnings freely and generously during the short time ashore between trips.
39. Interview with Jim, November 2012.
40. For example, see C. Lazenby and J. Lazenby, *Deep Sea Voices: Reflections of Women in our Fishing Communities* (Stroud: Tempus, 1999), p.28 and p.45.

Figure 3. Hull trawlermen at sea, 1955. Heavy gear on a moving deck and rough icy seas made distant-water trawling a hazardous occupation (Copyright Innes.co.uk).

respite, not long-term improvement, and had a role in maintaining fatalistic and short-term views amongst trawler crews.[41] Of greater concern was the contribution of fatigue to the poor safety record of the side trawler. As Piper observed, 'worn-out men are prone to blunder', and an error could have serious consequences.[42]

Fishermen in Hull and other distant-water ports were engaged on a casual basis, with employer or employee able to terminate engagement at the end of a trip without prior notice. Payment was made via a basic wage, supplemented by a share of the trip's profit, known as 'poundage'. Skippers and mates were paid largely on a net percentage of the sale of the catch. Pensions for fishermen were not introduced until 1961. Working hours could be exceptionally long and whilst poundage brought the potential

41. Tunstall, *The Fishermen*, p.131.
42. Piper, *North Ships*, p.87.

of high earnings, Tunstall suggests that in the late 1950s the hourly rate was 'probably lower than that of almost all adult male workers in Britain'.[43] Payment was calculated on the sale value of the fish and as this fluctuated, incomes could boom or bust. Behind the uncertainty and risk, however, lay the chance of a lucrative catch. Arguably, it was the call of the cash, not the sea that lured many young men into trawling.[44] To Tunstall this was particularly true for boys whose lack of academic tendencies made it difficult to achieve better-waged jobs ashore.[45] With trawling, if the long hours and risks were accepted, there was the chance of substantial reward. Big money brought local status and could inject excitement and anticipation into the hardships.

The casual work structure and dependency on poundage placed industry management firmly in power and fishermen in a position of non-negotiation. The history of capital and labour relations in trawling is heavily weighted in favour of management and remained so into the 1970s, a time when labour interests in other industries were defended by powerful unions.[46] In his Fabian publication Tunstall defined the capital–labour relationship as one based on 'mutual suspicion, and bitterness'.[47] Historians Mumby-Croft and Barnard have described the attitude of the trawler owners to their workforces as 'intransigent'.[48] Peter came to Hull to work for the WFA in 1965, having previously worked in the Merchant Navy. Observing working conditions and industrial relations within British trawling, Peter felt as though the clock had been turned back to the late nineteenth century. He describes the industry as 'brutal', saying 'whether it was Grimsby or Hull, there was a very strict pyramid structure and the gap between the top and the bottom was enormous'.[49] With no job security or protection, men committing indiscretion in the eyes of the management could be sent 'walkabout', putting them out of a ship for a period of time. Heavy pressure rested on the skipper, and a good fish-catching record could count for little should fortunes turn for the worse. In such an environment it was tempting to take risks. Within this structure, however, it is acknowledged

43. Tunstall, *The Fishermen*, p.69.
44. Horobin, 'Community and Occupation', pp.350–351; Tunstall, *The Fishermen*, p.109.
45. Tunstall, *The Fishermen*, pp.101–102.
46. See Mumby-Croft and Barnard, 'Antiquated Relationship?', p.120.
47. Tunstall, *Fish*, p.11.
48. Mumby-Croft and Barnard, 'Antiquated Relationship?', p.122.
49. Interview with Peter, April 2013.

that 'there were good trawler owners and there were not so good trawler owners': companies that showed concern for employees and those that did not.[50] Peter too concedes that 'some owners were better than others',[51] and oral history testaments often reveal the varying reputations held by Hull trawling companies.

Tunstall attributes the persistence of vulnerable working conditions to a fatalistic tradition amongst trawlermen and weak unionism.[52] As trade unionists, trawlermen were problematic. At sea for long periods, they were a fragmented group, difficult to contact, motivate and organise. Mumby-Croft and Barnard detect an 'anti-collectivist ethos' amongst fishermen, and Tunstall observed how unions could find themselves at ideological odds with their workforce.[53] Within larger bodies, such as the Transport and General Workers Union, trawling was a marginal and little-comprehended sector.[54] It was the trawler engineers, backed by a longstanding engineering tradition, who formed the backbone of fishing unionism. But here tensions between those 'down below' and the 'deckmen' disrupted any sense of unity.[55] In Hull, strike action was generally unofficial and reactive to short-term issues, rather than targeted at long-term improvements.[56] Tunstall writes, 'what happens on trawlers can only be altered ashore', and negotiating change required organisation, tact and forward planning, attributes that were not prevalent in trawling culture.[57]

Despite harsh and hazardous conditions, trawlermen could gain satisfaction from their extreme occupation. Trawling was not for everyone,[58] but many who embraced it remained firmly bound to its way of life. Victor reflects positively upon his career with the sidewinder fleet, feeling that given the chance, many of his peers would return to fishing as it had been

50. Interview with Ron, June 2013.
51. Interview with Peter.
52. Tunstall, *The Fishermen*, pp.172–174, see also Chapter 11.
53. Mumby-Croft and Barnard, 'Antiquated Relationship?', p.126; Tunstall, *The Fishermen*, pp.248–251.
54. Mumby-Croft and Barnard, 'Antiquated Relationship?', pp.124–125.
55. Tunstall, Fishermen, pp.247–248.
56. Tunstall, *The Fishermen*, p.250.
57. Tunstall, *The Fishermen*, p.134.
58. Within my sample of seafarers, two report that they decided not to continue with trawling. One suffered from seasickness and chose to work in the fishing industry ashore. The other alternated between the Merchant Navy and trawling, eventually choosing to come ashore to a bakery job, which he much preferred.

in their past.[59] The attraction was more than the big money. Poundage had its lure, but the job itself also held an appeal, with the strong work ethic of trawlermen both self-reported and widely noted by others. Bill, a trawling deckhand, appreciated the variety of tasks that were part of a typical trip – bringing the fish aboard, then gutting and transferring it from washer to fish room before storing it away. 'Everyday there was a different thing happened,' he says, 'you never knew what you was gonna do or what [*might*] come up next.'[60] Trawler mate Doug describes his own enjoyment of repetitious yet rewarding hard work, along with fresh air and the pleasure of calm after rough seas. 'When it fines away,' he says, '[*and*] the ship's steady, that in itself is an enjoyment … especially if the sun's out and it's not too cold and you're in your shirt sleeves … [*you feel*] oh it's good to be alive.' The attraction lay in a combination of the way of life and the job itself. 'It's repetitive,' Doug says. 'People who like gardening, going out to dig the allotment over – they'll get pleasure out of that – and it's the same with fishing.'[61] At the end of a trip, Jim remembers the satisfaction of steaming home. 'It might be fine weather,' he says, 'or a moonlit night – lovely – you're stood at the bridge window, everything's right, you got [*fish*] below – [*there's a*] lot of compensations.'[62]

The fulfilment from meeting the demands of a rigorous job is sometimes conveyed with a hint of 'otherness' and a distinction between trawlermen and other seafarers. Trawler engineer Dave describes carrying Royal Navy personnel aboard his vessel during the Cod Wars. Unaccustomed to the movement of a side trawler, the naval passengers suffered from seasickness. 'I came up to have a cup o' tea,' Dave recalls, laughing, 'and there was all three of 'em laid out in the mess deck. Sick as a dog. They asked me – how do you survive this? I just said – it's quite normal for us, y' know.'[63] A sense of difference, along with the shared rigours of an extreme occupation, resulted in a strong and enduring sense of camaraderie. To Victor, camaraderie with the people that he worked with was an important part of the job. 'You worked hard with them,' he says, 'you played hard. You trusted them and there was just something different about their whole attitude to life, that you didn't find anywhere else.' When asked if that was not the same in

59. Interview with Victor and Pauline.
60. Interview with Bill, July 2013.
61. Interview with Doug, March 2014.
62. Interview with Jim.
63. Interview with Dave, October 2013.

other seafaring jobs, Victor admits that there were similarities 'because you all went to sea', but in terms of ethos, he draws the conclusion, 'there was a difference'.[64]

Mumby-Croft and Barnard recognise a 'special industrial culture' within British distant-water trawling that can be traced to its late-nineteenth-century origins.[65] In the 1960s, that culture remained a complex and paradoxical brew. It included an absolute and non-negotiable management framework, but with a profit share system that meant everyone down to the trainee 'deckie learner' was invested in the pursuit of the big catch. It evoked the liberty of the hunter, yet was tethered by the principles of industry. It was characterised by common bonds, a sense of otherness and often a shared satisfaction in the job. But there were occupational tensions: fractured unions that at times did not understand their members and individualist and fatalistic men, hard to organise and disinterested in unions. Trawling was a 'culture of the moment'[66] forged by a hazardous and excessive environment, yet it commanded a workforce committed to their way of life. For a socialist and 'outsider' like Tunstall, the trawling industry presented the epitome of archaic practice and exploited labour. Yet amidst capital–labour tensions it is possible to detect in the workforce a cultural acceptance, even a complicity in the status quo. Reflecting on the industrial practice and culture of this extreme occupation, Mumby-Croft and Barnard conclude that whilst this 'invented world' may have appeared archaic from without, from within 'it had a coherent internal logic of its own'.[67]

A Place-centred Industry

> Fishermen see themselves in a curious way as working in Hull.
>
> J. Tunstall, *The Fishermen: The Sociology of an Extreme Occupation*

Prior to 1976, the distant-water fishery in Hull consisted of an entirely localised structure. Vessels, along with fish-landing and fish-processing facilities, were Hull owned and Hull based. Trawlers sailed from Hull and returned to the city with their catch. Added to this was a seagoing

64. Interview with Victor and Pauline.
65. Mumby-Croft and Barnard, 'Antiquated Relationship?', pp.125–126.
66. Mumby-Croft and Barnard, 'Antiquated Relationship?', p.126.
67. Mumby-Croft and Barnard, 'Antiquated Relationship?', p.126.

Figure 4. Overlooking the terraced houses of Harrow Street, Hessle Road, 1978 (Copyright Alec Gill).

workforce, where 92 per cent of men lived within four miles of the fish dock.[68] Men sailed predominantly with others from Hull and Tunstall observed a strong affiliation to the city.

Within Hull, fishing activity was concentrated within the district of Hessle Road. The popular notion of Hessle Road as a world of fishermen is well documented, but it has been subject to challenge. Tunstall, for example, used Horobin's 1955 figures to assert that only 10 per cent of the district's adult male population were fishermen.[69] Walton too draws

68. Horobin, 'Community and Occupation', p.345.
69. Tunstall, *The Fishermen*, p.85.

upon the same figures to demonstrate the beginnings of dispersal, showing one-quarter of fishermen living more than two miles from the fish dock.[70] Nevertheless, in 1955 two-thirds of fishermen did remain within a 1.5-mile radius of the dock and although a diverse urban economy ensured that the population of Hessle Road were not all fishermen, a significant number did live there. Horobin himself shared a more nuanced understanding of the relationship, in line with Walton's idea of a fishing community based upon 'commitment to an industry, not necessarily entailing actually going to sea, but being part of that network of shared interests and concerns that surrounded the fishing'.[71] It is a definition that includes dockworkers, fish processors, suppliers, retailers and publicans. To Horobin, the district of Hessle Road was one where everything was '"geared" to the rhythm of fishing', and in this sense Tunstall concedes that 'the fish-dock as a whole is easily the main employer of the district's men'.[72]

Tunstall recognised Hessle Road as a recruiting ground for fishermen. However, he contests the notion of 'fishing families', and the idea that trawling was a family tradition. Horobin and Tunstall both observed trawlermen actively discouraging their sons from entering a trawling life. Horobin, however, detects a crucial connection that is not fully appreciated by Tunstall. Establishing the strong link between industry and district, Horobin defines fishing as a *neighbourhood* tradition. He demonstrates that many fishermen were indigenous to the fishing district when they first entered the industry, and once at sea remained bound to the area that supported their unique way of living.[73] Like a mining community and its pit, there was a predisposing link between living on Hessle Road and working in the fishery, diffused only slightly by the wider urban context. Fishing became a 'family tradition' *because* of the neighbourhood. Here, young men lived in close proximity to other fishermen and, witnessing their apparent wealth and virility, chose to follow them to sea.[74] In the sense of being handed from father to son, fishing in Hull may not have been a true family tradition, but the bonds of place ensured that it ran in families.

70. J. Walton, 'Fishing Communities, 1850–1950', in Starkey et al. (eds), *The Commercial Sea Fisheries of England and Wales since 1300* (London: Chatham, 2000), p.131.
71. Horobin, 'Community and Occupation'; Walton, 'Fishing Communities', p.128.
72. Horobin, 'Community and Occupation', p.348; Tunstall, *The Fishermen*, p.131.
73. Horobin, 'Community and Occupation', pp.349–350.
74. Horobin, 'Community and Occupation', pp.349–350.

Fear, Strength and Getting On With It: Women's Experience of an Extreme Occupation

The practice and culture of the 21-day trip also impacted upon the lives of women ashore. Husbands, fathers, brothers and sons were absent for long periods, and when they were home it was in short and celebratory bursts. With men at sea, it fell to the women of the fishing community to provide the consistency that held families together. Wives managed households and paid the bills, whilst mothers raised children as virtual single parents. At a time when female independence was less prevalent, those with men at sea relied upon the help of family or joined together to provide essential networks for support and friendship.[75] Gill, wife of trawler skipper John and mother of four, recalls some of the difficulties. 'It was very hard,' she reflects, 'because you didn't have a man there. You couldn't just pop to the shop or say, oh I'll leave the kids with you – or on a night, when they came home, you couldn't go out the door or anything, you couldn't leave your children.'[76] For Gill, the support of her family was a lifeline. 'I had a good mam and dad' she says, 'and spent a lot of time with [*them*], when my husband was away.'[77] As well as providing support, friendships with other fishermen's wives brought company in what could otherwise be a lonely way of living. Gill recalls taking the children to the seaside with one such friendship group. Pauline, married to Victor, also remembers companionship and trips out. 'We'd ring each other up and say, when's your husband home? What's he made [*earned*]? When's he docking?,' she recalls, 'but we made a special effort when it was summer. We'd all meet up at East Park … and all our husbands was away. We put to and made picnic baskets and that sort o' thing.'[78]

In addition to the challenges of managing home and family alone, women lived with the constant nag of worry. Trawling was dangerous and men died. In interviews, male narrators rarely alluded to fear, but women with a trawlerman as husband, father or brother, regularly reported fear as a feature in their lives. Anxiety was not unfounded. Gill had lost her first

75. Lazenby and Starkey refer to trawling ports as a 'matrilocality'; C. Lazenby and D. J. Starkey, 'Altered Images: Representing the Trawling in the Late Twentieth Century', in Starkey et al. (eds), *The Commercial Sea Fisheries of England and Wales since 1300* (London: Chatham, 2000), p.171.

76. Interview with Gill, September 2013.

77. Interview with Gill.

78. Interview with Victor and Pauline.

husband and Pauline her brother-in-law in trawling incidents. Gill talks about her fears upon becoming a fisherman's wife again upon her second marriage. 'I married John,' she says, 'and I knew that he went to sea, but in the back of my mind, I always had the doubts – that this [*i.e. loss*] does happen to you – it doesn't only happen to others – and I was always quite afraid of it really.'[79] It was not necessary to experience a loss at first hand to feel its affect. Loss was a well-known spectre in the close-knit streets of Hessle Road. Lily, whose father was a chief trawler engineer, became aware of the dangers from an early age. 'We used to worry,' she recalls. 'That was the main thing … especially if something happened, y' know, to other ships. You thought, oh no, is it his turn next?'[80] There was no option but to relegate fear to the back of the mind. As Lily explains, 'You just coped and you just sort of got on with it really.' Pauline reflects 'it was just part of your life. You sort of accepted … well not accept, but hoped it never happened.'[81] Christine, married to trawler deckhand Paul, returns to the importance of female networks in dealing with worry. If her friends became concerned about their husbands or boyfriends at sea, they would find comfort together. 'We used to get a group and meet in town,' she recalls, 'and we used to talk about it.'[82]

After three weeks alone came the excitement of a homecoming. For the women as much as the men, this was a time for celebration. Pauline describes the anticipation. 'There was just that feeling of, oh God, he's coming home,' she says, 'and you'd get the house clean. Kids couldn't go in the front room and sit down. You'd get yourself ready.'[83] For women, the experience of having their men ashore could vary. Barbara, wife of a trawler deckhand, remembers some good times ashore. But she also recalls less agreeable occasions when crews would be 'out boozing with their mates'. 'By the time they come home,' she reflects, 'it was nearly time to go away again.'[84] Lily, however, remembers her father spending most of his time with the family, taking them out and giving them treats. A husband at home could be a time of companionship, celebration and sociability. Christine recalls, laughing, 'oh, it was good, very good. Yeah, we was

79. Interview with Gill.
80. Interview with Lily, September 2013.
81. Interview with Victor and Pauline.
82. Interview with Christine, September 2013.
83. Interview with Victor and Pauline.
84. Interview with Barbara, September 2013.

always out.'[85] Gill too reports that a night out was an enjoyable part of the shore-time experience. On settling day there would be drinks and maybe a meal with her husband's crewmates and their wives. It was a regular experience that she describes as being part of 'this community'.[86]

On the punishing cycle of the 21-day trip, time spent with loved ones was brief. Three days later, it was back to sea. Barbara poignantly describes her husband watching over their sleeping children as he waited to leave. He would sit for hours, she says, 'just staring at the little ones in their cots, y' know, til the taxi come'.[87] Gill defines departure day as difficult. 'I could never handle the going away,' she says. 'I used to go in silent mode and really, when I think about it, I used to get quite upset about it. But as the years went on, I learnt to handle it better.'[88] The rigours of the 21-day trip placed heavy demands upon the women at home, as it did upon the men at sea. Female strength is communicated overtly and implicitly within memory narratives of Hull's fishery, as it is within other seafaring communities. Hagmark-Cooper presents the figure of the strong 'sailor's wife' as a complex interplay between collective myth and individual reality. Nonetheless, it creates a powerful and productive emblem.[89] Smith has explained the role of internalised communal discourse in shaping behaviour and practices.[90] Just as at sea, a trawling life ashore did not suit everyone. But for the trawling community to succeed, its women must be robust, and the idea and performance of female strength were perpetuated and enabled by well-established matriarchal connections and a shared discourse of survival.

A Tragic Turning Point

Early in 1968, Hull received news that three of its distant-water trawlers had sunk within the space of just five weeks. The loss of the *St Romanus* (11 January, 20 men lost), the *Kingston Peridot* (26/27 January, 20 men lost) and the *Ross Cleveland* (5 February, 18 men lost), became an important catalyst for the British distant-water trawl industry. In March 1968, a Committee of Inquiry headed by Admiral Sir Deric Holland-Martin began an investigation into the safety of deep-sea trawlers and their crews. The

85. Interview with Christine.
86. Interview with Gill.
87. Interview with Barbara.
88. Interview with Gill.
89. Hagmark-Cooper, 'Sailor's Wife'.
90. Smith, *Uses of Heritage*, pp.14–15.

accompanying assessment of mortality rates among British fishermen, led by Professor Schilling, demonstrated that between 1959 and 1963 the risk of death from accidents at work for fishermen was 17 times greater than that of all men in England and Wales. Between the ages of 15 and 44 the risk rose to more than twenty times greater.[91]

The Holland-Martin inquiry considered the broadest spectrum of factors impacting upon the safety of a trawler. Alongside technical matters of equipment and vessel design, the practices and culture of the industry were opened to scrutiny. The ensuing report identified that a combination of long working hours, short time ashore, fatigue, contract insecurity and a competitive environment, conspired with the more obvious risks of seafaring to foster an atmosphere where loss and injury were accepted as a peril of the job. In particular, the report lamented that the industry as a whole (with certain 'marked exceptions') had failed to establish a satisfactory working relationship between workforce and management. In presenting its findings, the inquiry acknowledged the role of vessel improvements in increasing safety, but added:

> There are limits, however, to what can be done in terms of the design of ships and equipment to make the working environment on a trawler intrinsically safer. The problem is fundamentally one of changing human attitudes at every level of the industry, and of instilling greater awareness of the importance of safety in crews, skippers and senior management alike. Fatalistic attitudes towards accidents are a particular problem. Many fishermen regard the present accident rate as an inevitable consequence of their occupation.[92]

The inquiry was accompanied by a high-profile campaign for safety led by the woman of Hull's fishing community. In the past, women had been expected to endure the consequences of the industry, without having any influence over its practices. After 1968 this was no longer tolerated. Although controversial in some quarters of the fishing community, trawlermen had acquired a vehement proxy ashore. The findings of the Holland-Martin inquiry, along with increased media attention and the women's vocal campaign, became the spotlight that brought this little-understood sector

91. D. Holland-Martin, *Trawler Safety: Final Report of the Committee of Inquiry into Trawler Safety* (London: Committee of Inquiry into Trawler Safety, 1969), pp.118–121.
92. Holland-Martin, *Trawler Safety*, p.iv.

of seafaring into public consciousness. With it came a drive for change. Following the Holland-Martin report new practices slowly adjusted the operation of the British distant-water industry. April 1970 saw the launch of a 'mothership' or Trawler Support Vessel to support the fleet at sea. *MV Miranda* was managed and manned by the Hull shipping firm Ellerman Wilson Line, with a Master appointed by HM Coastguard. The vessel was positioned between Greenland and Iceland throughout the hazardous winter months.[93] *Miranda* carried a meteorologist, a doctor, electronic and radio officers, engineers and an ex-skipper fishing advisor. Their role was to monitor the circumstances of the distant-water fleet and to provide advice and direct assistance in the event of a problem.[94]

Although the triple trawler disaster was an important trigger in revising the operation and culture of British trawling, it was not the only factor. The Holland-Martin report had noted that the more progressive trawler owners had begun introducing improvements even before the circumstances that led to the inquiry. For example, some distant-water firms had introduced, or were considering introducing, one- or two-year employment contracts for skippers and mates. The report commended the efforts of Hull owners in introducing a fishermen's registration scheme, experimenting with training ships and attempts to set safety standards via the Hull Steam Trawlers Mutual Insurance and Protecting Company, albeit for the purpose of reducing insurance premiums.[95]

There was however, a greater transforming factor, and that was the stern freezer trawler. By 1956–57, the British distant-water industry was at its post-war pinnacle in terms of modernisation and landings.[96] From such a climax, performance began to fluctuate, with a pattern that was generally downward. Overfishing in the North-East Atlantic was beginning to take its toll, resulting in falling yields from a diminishing fish stock.[97] The fluctuating performance of the catching sector was accompanied by early signs of problems in the fish trades. The distant-water industry was founded on demand for inexpensive food, but in the affluent post-war years, consumer interest in fish declined. Whilst total supplies of all fish types

93. See http://nigelhadley.org.uk/page3.html (accessed on 17 May 2014). Nigel Hadley served as radio officer on *Miranda* between 1974 and 1980.

94. See http://nigelhadley.org.uk/page2.html (accessed on 17 May 2014).

95. Holland-Martin, *Trawler Safety*, pp.85–97.

96. Thor, *British Trawlers*, p.144; Robinson, *Trawling*, p.212.

97. Robinson, *Trawling*, p.214.

remained near pre-war levels, the growth in population meant consumption per capita had dropped.[98] Consumers chose to redirect their increased spending power to meat, poultry and other foodstuffs. The fish and chip shop prevailed, but in numbers gradually declining from its interwar heyday.[99] Frozen fish was yet to be accepted as a quality product and despite marketing campaigns by the British Trawler Federation and the WFA, and the gradual growth of processed fish-based convenience foods, the overall trend in fish consumption was downwards.[100]

By the late 1950s, poor catches, low prices and disputes over territorial waters were making distant-water trawling increasingly unprofitable.[101] In 1960, a government inquiry was launched into all aspects of Britain's fishing industry. Amongst the recommendations of the resulting Fleck Committee Report was the extension of government subsidies to the distant-water sector, targeted at reinstating the viability of the industry.[102] Bolstered by financial assistance, the British owners moved towards a technology that had the capacity to revolutionise the scale, intensity and geography of distant-water fishing.

Freezer Technology

British distant-water trawling had long been bound by the challenge of bringing a perishable produce back to a home market. The constraints of the 21-day trip meant the Hull skippers habitually focused on Iceland, Norway, Spitsbergen, Bear Island and the Barents Sea. Voyages to Greenland had revealed the limits of traditional vessels, with the long journey reducing the time spent fishing and extreme weather increasing the already considerable risk.[103] Robinson observes how major enterprise in British trawling typically arose in response to changes in market conditions or yield.[104] In Hull, pressure on North Sea fish stocks in the 1880s had stimulated the move to

98. Robinson, *Trawling*, p.213.
99. Fried fish shops in Britain reduced from between 20,000 and 30,000 in 1933 to 17,126 in 1950 and 12,000 in 1975. C. Reid, 'From Trawler to Table: The Fish Trades since the Late-Nineteenth Century', in Starkey et al., *England's Sea Fisheries*, p.161.
100. Reid, 'Trawler to Table', pp.162–163.
101. Robinson, *Trawling*, p.219.
102. Robinson, *Trawling* pp.219–220.
103. Tunstall, *The Fishermen*, pp.51–52; Robinson, 'Steam Power', p.214.
104. Robinson, 'Steam Power', p.216.

steam, enabling longer journeys to more distant grounds. In the 1960s, with greater fishing effort producing lower catches for a contracting market, the industry turned again to new technology: this time to the freezer trawler and its promise of efficiency and reward.

Small-scale experimentation with freezing fish is known from Denmark, Germany, the USA and Britain in the early decades of the twentieth century.[105] However, it was not until the 1950s that sustained progress towards the modern factory freezer trawler commenced. In 1954, the British company Salvesen redirected its experience of factory whaling ships to the fishing industry.[106] By adapting the stern ramp of the whaler, new stern trawlers were able to bring the catch aboard whilst steaming head to wind, eliminating the danger and effort of bringing the catch over the side. Adopting a factory template, men could process fish below decks, shielded from the Arctic weather. The technology of freezing at sea allowed vessels to sail further from the home port and to stay away for longer, spending eight to ten weeks at sea. The stern factory freezer was bigger, carried larger crews and had a greater capacity for catching and processing fish. With this new breed of ship came the biggest change in the practice and culture of the British distant-water industry since the adoption of steam.

In 1954, Salvesen led the way with the launch of their first stern factory freezer, the *Fairtry*. In Hull, the industry did not immediately embrace the new technology. In the latter stages of Hull's post-war fleet modernisation, the *Fairtry* had been an expensive and experimental prototype, and owners had adhered to the traditional wet fish side trawler. But in the early 1960s, with growing pressure on profit margins and with government assistance available, the circumstances warranted change. In 1961, Hull's first stern fishing trawler arrived in port. The *Lord Nelson*, built in Bremerhaven for Associated Fisheries, was a hybrid vessel landing a catch that was part fresh and part frozen at sea.[107] In 1962, Hull's first whole fish stern freezer trawler, the *Junella*, entered service for the trawler company Marr.[108] The evolution was complete in 1967, with the introduction of Hull's first factory freezer

105. I. Heidbrink, 'From Sail to Factory Freezer: Patterns of Technological Change', in D. J. Starkey and I. Heidbrink (eds), *A History of the North Atlantic Fisheries: From the 1850s to the Early Twenty-First Century* (Bremen: Verlag HM Hauschild, 2012), pp.62–63; Robinson, 'Steam Power', p.214.
106. Robinson, 'Steam Power', p.215.
107. Robinson, 'Steam Power', p.221.
108. Thompson, *A Tribute*, p.13.

trawler, *Coriolanus*, which was able to land fish that was both processed and frozen at sea.[109] By 1969, Britain possessed 33 freezer trawlers, 22 of which operated from Hull. The 22 freezers represented 20 per cent of Hull's total fleet of 93 trawlers.[110] The rest continued to land wet fish. By the 1970s, improvements in sonar enabled fishermen to locate and identify fish stocks, supplementing their dependency on experience and good fortune. The pelagic or mid-water trawl, introduced from the early 1970s and adopted by distant-water fleets, was a further break with the past. Combined with sonar technology, the mid-water trawl enabled fish to be taken from parts of the ocean previously shielded from fishing activity.[111] Primarily devised for pelagic species such as mackerel, the mid-water trawl could also capture species such as cod, haddock or blue whiting. In the unpredictable circumstances that were to emerge, the pelagic trawl would offer the flexibility to diversify and move between new and interchanging forms of fishing.

By the end of 1973, the Hull freezer fleet stood at 36 vessels.[112] Hull was rapidly emerging as Britain's leading distant-water freezer port, with a growing need for shoreside facilities that could accommodate new methods of landing and storing frozen fish. Traditional 'freshers' continued to land alongside freezer vessels, but from the mid-1960s, no further distant-water wet fish trawlers were built.[113] For Hull, the freezer was perceived to be the future.[114] However, whereas Britain had delayed the embrace of the freezer trawler until prompted by circumstances, other nations had been less reticent. In the early post-war decades, Eastern European countries had begun to build up their own fishing industries. By the 1960s, Russia, Poland and East Germany had acquired significant distant-water fishing fleets where the factory freezer featured prominently. Britain was losing advantage to the newly emerging competition.[115] More significantly, the technology of the freezer factory trawler, combined with the increasing efforts of new fishing nations, was causing alarm. Historian Heidbrink concludes: 'The crux of the matter was the threat posed by the highly efficient catching vessels of the European fishing nations to local fisheries,

109. Thompson, *A Tribute*, p.14.
110. Robinson, 'Steam Power', p.215; Thompson, *A Tribute*, p.14.
111. Heidbrink, 'Sail to Factory', pp.64–65.
112. Thompson, *Fish Dock*, p.37.
113. Robinson, 'Steam Power', p.215.
114. HCA (HC) Ref: C/DBTH/59/21. The UK Fishing Fleet in the Mid-Seventies. White Fish Authority, Fishery Economic Research Unit, August 1969.
115. Robinson, 'Steam Power', p.215.

Figure 5. A new breed of ship: the freezer trawler *Sir Fred Parkes*, built 1966. By the early 1970s, Hull was emerging as the leading UK freezer trawling port (Copyright Maritime Museum: Hull Museums).

particularly those of new sovereign nations like Iceland.'[116] As alarm turned into action, for distant-water trawling ports like Hull, the North Atlantic bonanza was poised to come to an end.

The Freezer Generation

In June 1972, a feasibility study undertaken to inform the proposed relocation of Hull's fishery from St Andrew's Fish Dock to the adjacent Albert and William Wright Docks outlines the composition of the Hull fleet.[117] It reveals an industry in transition: 66 wet fish trawlers, 24 freezer

116. Heidbrink, 'Sail to Factory', p.69.

117. HCA (HC) Ref: C/DBTH/15/129. HFVOA. Preliminary Feasibility Study of Proposed Re-location of Hull Fishing Fleet. G. Maunsell and Partners, June 1972, p.3.

trawlers, one part freezer, one factory ship and eight seine netters. The report predicts that by 1987 wet fish trawlers would be reduced to merely ten.[118] It goes on to estimate that by 1974, a further 11 freezers would have joined the fleet and by 1987 the Hull freezer fleet could stand at 60. These optimistic figures place the stern freezer trawler at the forefront of Hull's fishing future. Freezer trawlers brought a new way of operating. They were safer, with a lower death and accident rate than the sidewinder, and with working and living space that was more comfortable. Freezers generated new systems of landing, selling and distributing fish that helped to make the trawlerman's wage more predictable and stable. There was also a significant difference in the amount of time spent at sea. Without the imperative to preserve wet fish for market, the freezer trawler could travel further afield and stay longer at sea to maximise the catch – generally between eight and ten weeks. Longer sea-time was compensated with more time ashore, and freezer trawlermen typically enjoyed one day at home for every week spent away.

Trawling – Your Future is a recruitment film produced in 1974 by the Port of Hull Fishing Industry and the Hull Nautical College.[119] It is a window into the changing fleet and reforming practices in the post Holland-Martin era. Shot during the last operational days of St Andrew's Fish Dock, the film's narrative firmly consigns the side trawler to the past. The promoted fleet is one of stern and freezer trawlers – the vessels of the future – and the lives of new recruits embarking upon a trawling career reveal significant departures from the earlier experience. Training is rigorous and structured around classroom learning, simulation and supervised sea-time. 'They're organised now down the dock,' explains the film commentary. 'The recruitment and training office, they're a special body to look after these young lads.' With more time spent at sea, the commentary explains, comes a greater need to relax and get away from the job. Work is depicted as balanced, with space to enjoy an onboard cinema or play darts on the mess deck. Cabins are single or double. A washing machine suggests the comfort of domestic routines amidst the trawling schedule.

118. The authors of the report note the HFVOA's own calculations of four wet fish trawlers in 1987, but believe that the figure would be closer to ten.

119. *Trawling –Your Future* (Antony Barrier Productions, 1976) [Documentary Film: British Film Institute Ref: C-64335]. The BFI dates the film to 1976, but must be earlier (possibly 1974). The footage is filmed on a fully operational St Andrew's Dock, which was closed to trawlers in November 1975.

One major divergence from tradition lay not in the structure of the modernising fleet, but in the young men recruited into it. Until the 1970s, every skipper had started his career on the deck. Trawling did not include an officer class and part of its traditional appeal must be attributed to the fact that with hard work and aptitude, a young man with little formal education could start as a galley boy and make it to the top. With the technologies of the freezer fleet – the bigger ships, sophisticated electronics and the complexities of freezing at sea – came a more scientific approach to fishing. To lead this advanced fleet, the Holland-Martin report had recommended that the industry recruit 'able young men with higher educational qualifications than are usual at present'.[120] Importantly, this should be without prejudice to the 'chance of promotion amongst able deckhands'.[121] In response, *Trawling – Your Future* announces that alongside progression from the deck, there is now 'another way to the top'. The Hull Fishing Vessel Owners Association Cadetship Scheme, named 'Project 70s', introduced a staged selection process for this new breed of entrant. Upon completion, recruits would be assigned to a freezer trawler for further instruction. The trawler cadet was intended to supplement, not replace, the existing ship's complement, yet it was recognised that with the right onboard supervision, cadets could 'quickly become a valued and useful assistant to any skipper or trawler officer in the dispatch of his own duties and responsibilities'.[122]

Alongside the cadetship scheme there were new geographies of recruitment. In the localised industry of the 21-day trip, most fishermen had come from Hull, and from Hessle Road in particular. Former fish dock barrow boy and trawler pipe-fitter William explains that West Dock Avenue School, located close to the fish dock, had a role in recruiting young men to the industry. Former pupils had gained positions of responsibility on the dock, and if a vacancy arose, the school would be contacted and asked to send on boys who were interested. 'That's how I got my first job,' says William, 'as easy as that.'[123] In *Trawling – Your Future*, accents are detected from much further afield than West Dock Avenue: they are Geordie, Scottish and from the Midlands. 'Nowadays,' explains the commentary,

120. Holland-Martin, *Trawler Safety*, p.7.

121. Holland-Martin, *Trawler Safety*, p.98.

122. HCA (HC) Ref: C/DBTH/15/123. HFVOA. HFVOA Cadetship Scheme c.1970. Project 70s, 21 September 1970.

123. Interview with William, March 2013.

'almost as many entrants come from inland as from the Port of Hull.' Former trawler mate Doug taught in the Fishery Department of the Hull Nautical College throughout the 1970s. He remembers a different type of recruit coming through the doors. Traditionally, most recruits had been the sons of fishermen or those with family ties to trawling, who had an idea of what the job entailed. Doug describes the new kind of recruit in the 1970s as 'green hands'. They were perhaps 'unaware of what they were letting themselves in for,' he recalls. 'You had to be a certain class of person to stick to trawling.'[124]

The practices of the freezer fleet – the longer time ashore, the more formal career structure and the new means of selling frozen fish – offered greater stability for fishermen and brought trawling more in line with other seafaring sectors. Freezer trawlermen now had less reason to live by the fish dock. In the past, trawler mates had returned to oversee the sale of wet fish at auction, and on settling day, the crew had returned to the dock to collect their post-sale poundage. Spending just three days ashore had provided an incentive to reduce travel time by residing within the concentrated spaces of Hessle Road.[125] But the introduction of new rhythms began to relax old dependencies. For the first time in decades, life aboard a trawler was no longer the same as the old fishermen remembered, as advancing technology remoulded the experience of both sea and shore.

Working for the WFA, Peter remembers the psychological and lifestyle adjustment that was required to adapt to the freezer fleet.[126] The new patterns of working did not suit everyone. Adjustment was further complicated by the fact that the increasing prevalence of the freezer trawler had collided with the growing prospect of fishing limits, causing some to question their future in the industry. Doug left trawling as early at the mid-1960s. 'The stern trawlers were coming out,' he says, 'which meant being at sea for two or three months, which really didn't appeal to me.'[127] The 21-day trip had become deeply engrained in the local trawling psyche. 'I was a three-week man,' Doug laughs, 'and [*liked to*] get home periodically, blow the cobwebs away.' Jim, concerned by expanding restrictions, left fishing in the early 1970s without ever having sailed on a freezer trawler. Again, the

124. Interview with Doug, November 2012.
125. Horobin, 'Community and Occupation', pp.347–348.
126. Interview with Peter.
127. Interview with Doug.

three-month voyage of the freezer ship did not appeal. Jim reflects that it 'wasn't my cup o' tea at all', and others echo his sentiment.[128]

Many however, did make the transition. For some, like deckhand Bill, the shift only occurred when the sidewinder fleet was scrapped. 'I had to go in a freezer,' recalls Bill, 'and I hated it.'[129] Despite sailing on freezer vessels for four years, Bill maintains his preference for the three-week trip, where 'you knew you was gonna get home'. Others describe the freezers in a positive light. Skipper Bob enjoyed the better conditions on the stern trawlers. In the 1970s, he says, 'as the ships got better, the way of life got better and easier – not so dangerous'.[130] There was even better food. Other narrators similarly made the transition without concern. Thomas and Kevin are amongst the younger respondents. They entered the industry in the late 1960s and early 1970s and both worked happily between sidewinder and freezer. Kevin acknowledged the pressure that longer sea-time might have placed upon men with families. However, speaking of himself, he says, 'I was only a young kid, so it didn't bother me where I went.'[131] Thomas captures the experience. Working on a freezer, he says, was 'massively different'. He describes a changing relationship with the sea. Side trawlers were smaller ships and as fish were gutted on an open deck, the sea was close and constant. On the freezer vessels the presence of the water felt less immediate. 'They're a lot bigger ship,' Thomas explains. 'You're a lot higher … the only time you really see the sea is when it comes up the ramp.'[132] The fundamental hands-on nature of the work did not change. For Thomas, it still felt like fishing, but with the comfort of working in an enclosed factory space. 'It can be as rough as hell,' he says, 'and you won't know anything about it.'

As the freezers altered practices at sea, they also changed lives ashore. For the wives of freezer trawlermen, longer times at sea meant coping alone for longer spells. Margaret's brother Harold worked on freezer trawlers and this could be hard for his wife and children. 'You can't put the [*weeks away*] back into your life,' she reflects.[133] For families accustomed to the 21-day trip, the eight to twelve weeks of a freezer voyage could seem an eternity.

128. Interview with Jim.
129. Interview with Bill.
130. Interview with Bob, July 2013.
131. Interview with Kevin, August 2013.
132. Interview with Thomas, September 2013.
133. Interview with Margaret, September 2013.

However, with one day off for every week at sea, at the end of a trip men could be home for over a week and this brought advantages. Gill's husband John sailed on the freezers and she remembers the benefits. 'You had a little bit more time to do things,' she says. 'He would still come home and you would still have that settling day, and you would still go out and we'd always take the children the next day for a meal. But then we had time together. We could plan holidays. We could go away a week.'[134]

Prior to the final Cod War, few reported that the impact of the freezer fleet could be felt on Hessle Road itself. There was the appearance of cold stores required to distribute frozen fish. There was perhaps a slight rein on the celebration and free spending, as the 'three-day millionaires' stayed at home for a week. From the early 1960s, Hessle Road was entering an era of adjustment unconnected with fishing and this may have rendered other change less visible. Eventually, the continued expansion of the freezer fleet would have remoulded the fishing district. In 1976, however, the new rhythms at sea were yet to fully dominate ashore.

By the start of the 1970s, the emergence of the stern freezer ship heralded the greatest adjustments to British trawling practice and culture since the arrival of steam almost a century earlier. Oral history narratives in Hull reveal the beginnings of a generational shift that saw some transition from sidewinder to freezer whilst others chose to avoid them. However, a new generation of trawlermen was emerging that viewed the freezer trawler as home. Trawler engineer Dave was one of a new breed. Entering the industry in 1968, his fishing career took place exclusively aboard freezer vessels. To Dave, these ships offered a more stable seafaring life, and when asked how he would like to commemorate the industry, he replies that it is a freezer trawler that he would put on public display.[135] But as Hull's freezer capacity was growing, a political storm was gathering that would have devastating consequences. In 1975, Hull trawlers sailed into a third and final Cod War with Iceland. The Hull fleet that faced exclusion from the high seas of the world was a hybrid of sidewinder and freezer vessel. It was a fleet of two personalities and two cultural rhythms. Janus-faced, it still nodded to the past as it reconfigured for the trawling future. The outcome of the final Cod War, however, meant that for Hull, a trawling future was not to be.

134. Interview with Gill.

135. Interview with Dave.

Chapter Three

A New Fishing World Order

From the 1960s, the advancing technology of a growing global fleet of large freezer trawlers accelerated the problem of overfishing. Overfishing was not new to the late twentieth century. Since the intensification of trawling as a fish-catching method a century earlier, policy makers, scientists and fishing communities had been caught in a seemingly irresolvable dispute over the bounty of the sea.[1] The discourse drew social, economic and environmental voices, each seeking to secure and serve their position. Over previous decades the voices had been audible at varying volumes and heeded to differing degrees. But in the post-war North Atlantic, the environmental tone was increasingly persistent, until, coupled with the changing language of territorial waters, it could no longer be ignored.

The link between intensive fishing effort, technological efficiency and the push to extend fishing limits is well established by historians and contemporaries.[2] Between 1945 and 1977, the drive to protect and control lucrative fish stocks would result in the breakdown of long-accepted wisdoms regarding the freedom of the seas and the collapse of established fisheries rooted in that premise. British distant-water fishing was anchored on the certainty that a nation's territorial waters were defined by a limit set three miles from the shore. This was a widely, although not universally, accepted convention, but to British minds it was enshrined in international law.[3] After the Second World War, the position was beginning to change.

1. A. Burton and J. Ramster, 'Government, Science and the Fisheries in the Twentieth Century', in Starkey et al., *England's Sea Fisheries*; J. Hubbard, 'Changing Regimes: Governments, Scientists and Fishermen and the Construction of Fishing Policies in the North Atlantic, 1850–2010', in Starkey and Heidbrink, *History of North Atlantic Fisheries*.
2. A. Gilchrist, *Cod Wars and How to Lose Them* (Edinburgh: Q Press, 1978), pp.52–56; Thor, *British Trawlers*; Robinson, *Trawling*, pp.224–231; Heidbrink, 'Sail to Factory', p.69.
3. Robinson, *Trawling*, p.224.

In 1945, the Truman Proclamations claimed US American control of the natural resources on the continental shelf off the US coastline, along with the right to establish conservation zones for coastal fisheries. By 1948, this had been echoed by Latin American countries, who claimed jurisdiction over their own 200-mile coastal zones.[4] The ripples soon reached British shores. Since 1936, Britain had contested the legality of a Danish proposal to claim a four-mile coastal territory. The claim was significant, for it changed the premise upon which territorial waters were calculated. Rather than following the usual line of the low watermark, the Danes proposed limits drawn from straight baselines stretching from various coastal features, some standing at a distance from the shore. The new system drew large tracts of coastline and fish-rich bays into territorial waters. In 1951, Denmark won an international court ruling that this approach did not breach international law. It was a turning point for the long-term prospects of British trawling interests, for as historian Robinson observes, Iceland had been viewing the Danish outcome 'with far more than academic interest'.[5]

In 1948, Iceland had passed a 'Fundamental Conservation Law', adopting the principle, although not yet the practice, of the right to regulate fisheries within the area of the continental shelf.[6] In 1952, with Icelandic waters under increasing pressure from foreign fishing fleets, Iceland seized the opportunity to follow the Danes in declaring a four-mile limit drawn from baselines. Britain protested at what it perceived to be a unilateral violation of the freedom of the high seas. In Hull and Grimsby, the British Trawler Federation imposed a ban on Icelandic landings. Iceland responded by establishing new overseas markets, including significant trade with Russia.[7] The move upset NATO members, keen to retain Iceland's cooperation and its strategic location in Cold War military logistics.[8] The stalemate was not broken until 1956, when Britain responded to negotiations by accepting the new limit, whilst maintaining the challenge to its legal validity.[9]

Icelandic historian Thor suggests that to many Icelanders, the four-mile limit was a first step on a journey towards 'reclaiming what they considered

4. N. Ashcroft, 'The Diminishing Commons: Politics, War and Territorial Waters in the Twentieth Century', in Starkey et al., *England's Sea Fisheries*, pp.224–225.
5. Robinson, *Trawling*, p.226.
6. Ashcroft, 'Diminishing Commons', p.225.
7. Robinson, *Trawling*, p.229.
8. Gilchrist, *Cod Wars*, p.63.
9. Robinson, *Trawling*, p.229.

to be the country's former rights'.[10] In 1958, the first UN Conference on the Law of the Sea failed to reach a definitive conclusion on the growing issue of territorial limits. In the face of indecision and spurred by falling catches, Iceland acted by unilaterally declaring a 12-mile limit operational from 1 September 1958.[11] As other nations withdrew from Icelandic waters, Britain again protested. Whilst diplomatic negotiations faltered, the British trawler fleet continued to fish in the disputed waters, protected from the Icelandic Coastguard by the Royal Navy. The first Cod War was underway. This initial seaborne conflict was a war of harassment, rammings and collisions between gunboats and trawlers. The fleet fished in 'boxes' where they could be protected by the Royal Navy. However, it was clear that the 12-mile limit could no longer be contested as contrary to international law. In March 1961, Britain conceded to Iceland's 12-mile limit. Where Iceland led, Norway, Denmark and Faroe followed and by 1964, 38 countries had introduced similar restrictions.[12]

By the early 1970s, international voices were increasingly speaking in terms of 200-mile limits. In July 1971, the new left-wing Icelandic government declared its intent to push fishing limits to 50 miles, effective from 1 September 1972.[13] Britain once again protested, this time actively supported by West Germany. British and German vessels continued to fish in Icelandic waters in what was to become the second Cod War. The second Cod War took a similar path to the first, although incidents became more violent and high risk. The Icelandic introduction of the trawl wire cutter, designed to sever the wires attached to a working trawl net, proved effective, costly and potentially dangerous.[14] Royal Navy intervention was largely ineffective. Back on land, in September 1973, Iceland's threat to break off diplomatic relations with Britain brought both sides to the table.[15] In November, a compromise agreement allowed British trawlers restricted access within the 50-mile limit for a two-year period. One month later, the third UN Conference on the Law of the Sea demonstrated widespread international support for the 200-mile limit. Even before the two-year agreement with Britain had expired, Iceland, arguing that cod stocks were

10. Thor, *British Trawlers*, p.170.
11. Thor, *British Trawlers*, p.170.
12. Ashcroft, 'Diminishing Commons', p.226.
13. Thor, *British Trawlers*, p.200.
14. Thor, *British Trawlers*, pp.204–208.
15. Robinson, *Trawling*, pp.239–240.

in a critical decline, extended its fishing limits to 200 miles, with force, from October 1975.[16]

Under the standing agreement, this did not apply to the British fleet until November.[17] However, British and German trawlers continued to fish in the disputed waters beyond this deadline, in defiance of Icelandic claims to jurisdiction. This third Cod War deployed the same tactics as the prior two, but with even more vigour. As before, the conflict at sea achieved nothing. In February 1976, following a series of unsuccessful talks, Iceland broke off diplomatic relations with Britain.[18] Increasingly isolated and in conflict with a fellow NATO nation, Britain drew upon her NATO allies to help resume negotiations. From June 1976, the Oslo Agreement allowed 24 British distant-water trawlers to fish in Icelandic waters at any one time. The agreement ran for just six months. When it expired on 1 December 1976, British trawlers left Icelandic waters for good.[19]

Authors have reflected upon the motives and outcomes of events surrounding the Cod Wars. NATO and the strategic significance of the US airbase at Keflavik is widely recognised as a fundamental influence. Johannesson suggests that Iceland's key position in Cold War politics was valuable not only to NATO, but also to Scandinavian countries, who benefited from the US presence without hosting them on home soil.[20] NATO would not risk losing an important tactical partnership over fishing rights, and knowing this, Gilchrist believes that Iceland played upon the 'fear of going to the Russians' with aplomb.[21] From an Icelandic perspective, fish was central to their economy and the principle means of earning foreign currency.[22] During the Second World War, Iceland, along with the Faroese, had built up significant capital reserves by supplying fish to overseas markets, including Britain. The profit was ploughed into developing their own fishing industries.[23] For Iceland, the post-war years

16. Thor, *British Trawlers*, p.218.
17. Robinson, *Trawling*, p.241.
18. Robinson, *Trawling*, p.242.
19. Robinson, *Trawling*, pp.242–243.
20. G. Johannesson, *Sympathy and Self-Interest: Norway and the Anglo-Icelandic Cod Wars* (Oslo: Institutt for Forvarsstudier, 2005), p.143 and p.46.
21. Gilchrist, *Cod Wars*, p.72.
22. Robinson, *Trawling*, p.226.
23. Gilchrist, *Cod Wars*, p.42; G. Johannesson, 'Life is Salt Fish: The Fisheries of the Mid-Atlantic Islands in the Twentieth Century', in Starkey and Heidbrink, *History of North Atlantic Fisheries*, p.282.

brought the promise of fish-based wealth, but the spectre of overfishing threatened to take this away, first with the demise of herring, then with warnings of cod depletion.[24]

Conservation, in parallel with economic interest, lay at the heart of Iceland's drive to control its waters. Writing in 1969, J. S. Griffiths offers this insight:

> Poverty the Icelanders understood and had endured … But to obtain undreamed-of wealth in less than a decade and then to see the threat of its abrupt withdrawal as other fishing fleets returned to the seas round the coasts of their country – this was more than the Icelanders could be expected to bear calmly and purely rationally.[25]

The push for limits can thus be seen as the struggle for control by a small, newly independent nation and developing economy. Once set upon this path the Icelandic approach was determined and sustained. Forging just ahead of international opinion, Iceland became the stronger nation; the 'icebreaker' that cleared the way for others.[26]

For the British, the Cod Wars have been cast amongst pivotal events marking a shift in Britain's position on the world stage and as a naval and maritime power. Having suffered defeat over the four and 12-mile limits, and in the face of a clear shift in international opinion, it is difficult to see why Britain continued to pursue defiance over negotiation. The British distant-water trawl industry had been built upon the concept of the freedom of the seas and the later Cod Wars could be seen as a tactic of delay in the hope of achieving an outcome that retained at least some familiar principles.[27] Johannesson has suggested that cultural factors may have played a role. He identifies the 'extraordinary psychological effort' that would be required for an old naval power like Britain to concede to maritime decisions taken in small places such as Reykjavik.[28] The Icelanders were perhaps not easy to bargain with.[29] Yet a fundamental factor for the British was their belief

24. Thor, *British Trawlers*, pp.232–235.
25. J. Griffiths, *Modern Iceland*, London, 1969, quoted in Gilchrist, *Cod Wars*, p.108.
26. G. Johannesson, *Troubled Waters: Cod War, Fishing Disputes and Britain's Fight for the Freedom of the High Seas*, Studia Atlantica: 11 (Hull: NAFHA, 2007), p.297.
27. Gilchrist, *Cod Wars*, p.110; Johannesson, *Troubled Waters*, p.303.
28. Johannesson, *Troubled Waters*, p.300.
29. Gilchrist, *Cod Wars*; Johannesson, *Troubled Waters*, p.297.

that they were right. Britain's position was argued throughout on what were perceived to be points of law, historical precedence and an adherence to the notion of the high seas.[30] But the nation was swimming against the tide. Gilchrist observes that early in the debate the tone of emphasis had shifted from law towards ideas of equity.[31] Britain's legalistic tenor was out of tune with international harmony. Change had become inevitable.

Whilst locked into conflict with Iceland, Britain and its fishing industry was simultaneously fighting on another front. In 1970, Britain began negotiations to join the European Economic Community (EEC), along with Ireland, Denmark and Norway. In anticipation of welcoming four established fishing nations with extensive, rich coastlines, the existing 'Six' EEC member states had finalised the terms of a new Common Fisheries Policy (CFP). The policy embraced all aspects of fishing from industry structure to market organisation.[32] At its core lay the contentious principle of equality of access for all Community vessels within the waters of all member states.[33] The introduction of the new policy on the very day that membership negotiations began had angered the applicant states, who considered it 'a sharp practice on the part of the Six, guaranteeing them access to the fish-rich waters, without being obliged to take the interests of the coastal states concerned into account'.[34]

It was a requirement that the applicants joining the EEC accepted existing Community policy and this now included the new CFP.[35] Nevertheless, it became clear that for the fishing policy to be acceptable to an expanded Community, some modification would be required. The problem of reconciling the CFP with national expectations led to Norway's rejection of membership in 1972. For Britain, however, the Heath government were determined that the fisheries problem should not stand in the way of entry into Europe.[36] Following the negotiation of some concessions, the CFP was accepted. The most notable compromise was a

30. Gilchrist, *Cod Wars*, p.60 and p.109.
31. Gilchrist, *Cod Wars*, p.60.
32. M. Wise, *The Common Fisheries Policy of the European Community* (London: Methuen, 1984), p.107.
33. M. Shackleton, *The Politics of Fishing in Britain and France* (Aldershot: Gower, 1986), p.98.
34. Wise, *Common Fisheries*, p.108; Shackleton, *Politics of Fishing*, pp.97–99. Shackleton (p.97) suggests that the policy was agreed the day before negotiations.
35. Wise, *Common Fisheries*, p.108.
36. Robinson, *Trawling*, pp.236–237.

deferral until December 1982 of the right to 'fish up to the beaches' of member states. In the interim, it was agreed that six-mile limits could be retained, which in special circumstances could be extended to 12 miles, providing historic rights were acknowledged. With 95 per cent of its coastline qualifying for a 12-mile zone, the British government claimed a victory. Others, however, were concerned by the temporary nature of the concession.[37] The agreement was an uneasy truce and it was not long before it ran into difficulty.

Britain became a member of the EEC on 1 January 1973. At this time the nation's distant-water fishery was still embroiled in the discord of the second Icelandic Cod War. Symes observes that the formulation of the European CFP in the early 1970s occurred when national sovereignty over fisheries applied only to narrowly defined territorial waters.[38] The Icelandic victory in the third Cod War, however, forced the challenges and opportunities of the 200-mile Exclusive Economic Zone (EEZ) onto the European agenda. As a result, the formation of a European fishery took place within unfamiliar and rapidly changing parameters. Symes speculates that member states may have been less willing to cede their sovereign rights had 200-mile limits been a long-established protocol.[39]

By the mid-1970s, fear that the spread of 200-mile limits would divert Community and non-Community fishing effort into EEC waters brought a pressing need for the CFP to be revised.[40] Difficult negotiations lay ahead and in the face of Community division, the Hague Resolution of 1976 was drawn up as a guideline for the future development of the policy. The Resolution confirmed the Community's authority over third country agreements, both for non-Community members fishing within Community waters and for Community members fishing in third county waters.[41] Most significant amongst the Hague Resolution proposals was the agreement that member states would 'in concert extend the limits of their national fishing zones to 200 miles'.[42] The result would be a 'community pond', subject to the jurisdiction of each member state, but remaining bound by the

37. Robinson, *Trawling*, p.237.
38. D. Symes, 'The European Community's Common Fisheries Policy', *Ocean & Coastal Management*, 35:2–3 (1997), p.139.
39. Symes, 'Fisheries Policy', p.139.
40. Wise, *Common Fisheries*, p.150.
41. Wise, *Common Fisheries*, p.158.
42. Wise, *Common Fisheries*, quoting from the Hague Resolutions, cited p.157.

agreements of the CFP, including the principle of equal access. The zones were established simultaneously by member states on 1 January 1977, but still to be resolved was the contentious issue of how such a pond should be shared and managed.

For Britain, the EEC's move towards the 200-mile limit presented a predicament. Until late in 1976, Britain continued to challenge the Icelandic stance through the third and final Cod War. At the same time, discussions were progressing on a collective EEC 200-mile zone that would include British waters. Shackleton argues that the consensus on a European limit reached at The Hague in October 1976 marked the end of Britain's ability to challenge Iceland's territorial rights.[43] It was only a month after British trawlers left the disputed Icelandic waters on 1 December 1976 that Britain itself, with apparent duplicity, established a 200-mile exclusion zone. The zone brought little consolation to the British distant-water industry, which had swallowed the bitterest pill. Whilst losing the right to fish in the waters of other nations, under the terms of the CFP they were forced to share the newly established British coastal territory with the EEC fishing fleet. If Britain felt 'doubly penalised', it was not without good cause.[44] Wise highlights the sum of Britain's potential fishing losses from this dual blow. With the loss of foreign grounds, distant-water ports like Hull and Grimsby faced a catch reduction of 300,000 tonnes.[45] Without the obligations of the CFP, the rich fish stocks within Britain's 200-mile limit could help offset losses. Yet instead, Britain was to contribute an estimated 60 per cent of its fish resources to a European community pond. Traditionally, British fishing in the waters of other member states totalled less than 0.3 per cent.[46] Of all Community nations, Britain clearly had much to lose in the negotiation of the CFP revisions.

From this back-foot position Britain had much to gain from the rigorous pursuit of a satisfactory outcome. The original CFP had elided the need for conservation-based decision making, but the pressures apparent by the mid-1970s meant that the revised policy would be based around Total Allowable Catch (TAC) for the maintenance of sustainable as well as profitable seas. The identification of the TAC, its conversion and distribution as national quotas, access to territorial waters and issues of enforcement

43. Shackleton, *Politics of Fishing*, p.106.
44. Shackleton, *Politics of Fishing*, p.106.
45. Wise, *Common Fisheries*, pp.146–148.
46. Wise, *Common Fisheries*, pp.146–148.

became the cornerstones of negotiation.[47] Driven by the influential lobby of the British Fishing Federation (BFF), Britain campaigned robustly for a best-case scenario. However, time and again British desires clashed with those of other members and with the principles of the CFP. The revision of the CFP was proving impossibly complex and Britain's persistent objections were contributing to the delay of a much-needed outcome.

The election of new Conservative British government in 1979 brought a new degree of flexibility and a staged approach to the CFP.[48] The early years of the 1980s saw incremental progress that began with an agreement on conservation measures and moved towards a structural policy, measures to deal with excess capacity and support for new fishing opportunities, culminating in an outcome on the thorny issues of access and quotas.[49] With the deadline for the expiry of the existing CFP looming, the stalemate had finally been broken. The revised CFP was adopted in January 1983. British gains included 37.3 per cent of the available fish stocks in European waters and a 12-mile national exclusion limit, subject to certain historic rights.[50] Inevitably, the outcome was a compromise, arguably giving cause for neither celebration nor despair. However, the policy had been almost seven years in the making. In Britain, and for Hull in particular, those years spanned a critical period of disruption in its distant-water fishing industry. The protracted wrangling over the terms of the CFP revision, often instigated by Britain itself, created governmental and administrative uncertainty that hampered operational attempts to repurpose the former distant-water fleets. By the time the CFP was available to direct and manage transition, for many in the British distant-water fishery it was simply too late.

The expansion of limits and entry into the EEC presented a critical turning point in the fortunes of Britain's distant-water trawl fishery. For almost a century the industry had enjoyed steady progress. From the 1960s, early signs of difficulty were met with government support and the beginnings of a new and advanced fleet. In Hull, as the industry entered the 1970s, despite evidence of overfishing and the wider problems of the national economy, the outlook was generally optimistic. Yet as the decade progressed and economic strains deepened, the fishery struggled against growing impediment. On top of this came the Cod Wars and the uncertainty and

47. Wise, *Common Fisheries*, pp.162–163.
48. Wise, *Common Fisheries*, p.204.
49. Wise, *Common Fisheries*, pp.204–243.
50. Shackleton, *Politics of Fishing*, p.113.

restrictions of a European fishery. By the mid-1970s, adverse conditions had combined to produce the perfect storm, and a decade that had begun with confidence descended first into turbulence and then into crisis.

Losing the Battle

> They sent Naval vessels didn't they and things like that and *[we thought]* we could force our will on another nation. And I think the world had changed.
>
> Chris, Hull Fish Merchant

Despite the expansion of international limits from 1945 and the long-running friction with Iceland, there is little evidence that the British distant-water industry had fully grasped the extent or pace with which their world would be transformed. Looking back to the 1970s, narrators offer a spectrum of remembered attitudes and expectations that they blend with their present perspectives. Concerned by the spread of territorial waters, some trawlermen had left the industry early. Jim, who came ashore in 1972, says that with Cod Wars and limits, he anticipated a 'fading into the sunset'.[51] After failing to settle into factory work, Jim eventually returned to sea as a coxswain for the Humber Pilots. Ken was a skipper with the freezer fleet when he too predicted difficulties. Ken had estimated that the 50-mile limit alone would lead to exclusion from some of the best fishing grounds and in 1973, he decided to accept a cartographer's job with the WFA.[52] One of his first tasks was to draw the new territorial limits onto charts of Icelandic waters. In taking new roles, both Jim and Ken sacrificed the substantial earning potential of the trawler skipper. However, their foresight brought them job security as the distant-water industry plunged into uncertainty.

On reflection, there is a certain acceptance of the Icelandic stance. David, a fish merchant, concedes that the industry was wasteful. Overfishing was encouraged by the poundage payment system and he recalls the overzealous withdrawal of fish from sale to keep the market buoyant. It was, he says, 'a sheer waste of fish' and he wasn't surprised that Iceland decided to take control of their waters.[53] Trawlermen were aware of the pressure on stocks.

51. Interview with Jim.
52. Interview with Ken, May 2013.
53. Interview with David, March 2013.

To Victor the limits were unpalatable, but not unreasonable. 'I think most sensible people realised that it was the only option,' he reflects. 'If it was a free for all, then fish stocks were just going to disappear.'[54] Without restrictions, the heightened technology of fishing would take its toll. Deckhand Thomas draws the blunt conclusion: 'We was catching miles too much fish.'[55]

However, empathy with Iceland could only be in principle. In practice, Hull's fishery depended upon distant waters. Working for the WFA, Peter felt that the distant coasts of the North-East Atlantic had come to be viewed as 'our waters'. Trawler skippers Jim and Ron both point to the pioneering role of Humber trawlermen in developing North Atlantic fishing.[56] From Newfoundland to the Barents Sea, Hull skippers had their favourite fishing grounds, which had become familiar through relentless cycles of the 21-day trip. If as Tunstall suggests, fishermen did indeed see themselves as working in Hull, then distant waters could easily become an extension of the city. It was a fallacy stemming from decades of journeys that had started and ended in the home port. Ron recalls the difficulty of accepting the new circumstances. 'We didn't think it was possible … that one country could suddenly declare that amount of seas within their limits,' he reflects. 'It was unbelievable that someone could just say well, this chunk of water is ours and nobody else is going to come into it.'[57] Extensive limits were a new phenomenon, and culturally as well as economically it required adjustment. In the late 1950s, Tunstall had remarked that Hull fishermen viewed life on a trawler as permanent and ordained.[58] In the 1970s, the same sentiment can be observed in attitudes to the industry itself. Ivy had worked as a fish processor and was born and raised on Hessle Road. To Ivy it was inconceivable that trawling should come to an end. Remembering the Cod Wars, she remarks:

> I did think we would [*win*] them … we should have won, because it was our lifeline. I don't see why we did lose it. All the people that went out, lost their jobs, they worked really hard. There was a lot of lives lost and we should've really won … We thought we'd have it forever.[59]

54. Interview with Victor and Pauline.
55. Interview with Thomas.
56. Interviews with Peter, Jim and Ron.
57. Interview with Ron.
58. Tunstall, *The Fishermen*, pp.133–134.
59. Interview with Ivy, April 2013.

Ivy's words reveal the taken-for-granted certainty of life in a long-established fishery. Politics aside, Ivy reveals the crux of communal concerns. This is our lifeworld and our livelihood. We work hard and men have died. From this perspective, it is possible to see how the North-East Atlantic came to be viewed as 'our waters'.

Similar sentiments no doubt impelled the men of the trawling fleet as they engaged in the Cod Wars. Trawling practice and culture left trawlermen ill-equipped for negotiation ashore, but in the Cod Wars, protest took place in an arena that they understood. Whilst trawler owners, via the BFF, sought to influence political decision making, trawlermen fought to influence outcomes at sea. Narrators who fished through the Cod Wars tell tales of evasive antics, costly arrests and near misses.[60] Most concede that the risks taken on all sides were dangerous. Lily worked as a secretary for the WFA on St Andrew's dock. She remembers the Cod Wars as a time of tension. 'It was dangerous for the men going out to sea,' she recalls, 'because they took chances ... so that was worrying for families. It was worrying for people who worked with them.' Lily projects the sense of 'otherness' that is so often discernible within trawling narratives. 'It was a bit odd really,' she reflects. 'It wasn't like having a war as such, it was just a certain group of people who were put in a firing line.'[61]

With livelihoods at stake, there seemed little choice but to stick it out at sea. Among seagoing narrators were those who, like Ivy, believed that the Cod Wars might be won. Deckhand John K. did not anticipate failure. Iceland was a small nation, and Britain had sent in naval vessels. 'We thought we'd be ok,' John says, 'but ... we just wasn't.'[62] Alongside optimism, by 1975 there coexisted an awareness that this was a last stand. Within the distant-water industry, owners, employees and community alike were forced to put their trust in politicians. Many ultimately felt let down. Engineer Dave had not anticipated victory in the third Cod War, but had held on to hope over conviction. 'We felt that our government was weak and they wouldn't put up a fight,' he says, 'and they didn't ... we hoped, but we all felt that we would lose the battle.'[63] Michael shares

60. Not all active trawlermen were involved in the Cod Wars. For example, deckhands Kevin and Bill sailed with skippers who favoured the White Sea grounds and therefore avoided the conflict.

61. Interview with Lily.

62. Interview with John, July 2013.

63. Interview with Dave.

tales of Cod War tactics, including his own arrest for breaching Icelandic limits. He too expresses disappointment, concluding '[*trawlermen*] thought the government had let us down big time, y' know, really let us down … we honestly thought that if we got stuck in, the government would back us up and they never.'[64]

Betrayal was also felt in the senior ranks of the industry. Tom Boyd was the third generation of Thomas Boyds to run the Hull trawling company Boyd Line. In a world of pending fishing limits, Tom had hoped that the progress of restrictions could be delayed and that when they finally occurred, British politicians would be 'sufficiently on the ball' to negotiate continued fishing access.[65] As a member of the BFF, Tom was also engaged in the simultaneous battle for a favourable Commons Fisheries Policy. The outcome was a double let-down. In Tom's opinion, Prime Minister Edward Heath was simply unwilling to consider the importance of the British fishing industry.[66] Hull, along with other trawling ports, was a safe Labour seat and entry to the Common Market took political precedence. 'It was just of no consequence to the government of the day that the industry should be sacrificed,' Tom reflects, 'and sacrificed it was.'[67]

Where change had been anticipated, most did not expect the pace with which their livelihoods would plunge into crisis, with little opportunity for gradual adaptation. In the wake of defeat, Victor echoes others when he says, 'most of us thought that some agreement would be reached between our government and the Icelandic government, with regard to continuing to fish there.'[68] But under the terms of the interim CFP, negotiating access to the newly expanded waters of non-member states was not directly in the hands of the British government. In the aftermath of the final Cod War, the reorientation of the British distant-water industry was to present a major challenge, and the certainties of trawling could no longer be taken for granted.

64. Interview with Michael.
65. Interview with Thomas (Tom) Boyd, November 2013.
66. During a research interview by Robert Saunders, Labour MP Jim Sillars similarly expressed his belief that fishing had been viewed as a 'throw away card' during EEC negotiations. R. Saunders, *Yes to Europe* (Cambridge: Cambridge University Press, 2018), p.296.
67. Interview with Tom Boyd.
68. Interview with Victor and Pauline.

Unprepared for the Future

In 1976, it can be argued that all sectors of the fishery were ill prepared for the trials ahead. In the hope for a best-case outcome, less attention had been paid to worst-case scenarios. The 1969 WFA report 'The UK Fishing Fleet in the Mid-Seventies' predicts a decline in catch from all North Atlantic waters, resulting from a growing international fishing effort and reducing fish stocks.[69]However, it makes little reference to Iceland and there is no evidence that the imminent and destructive potential of EEZs had been fully considered. Similarly, the 1972 feasibility report exploring the future relocation of Hull's trawling fleet from St Andrew's to the adjacent Albert and William Wright Docks fails to accommodate any sense of threat.[70] Plans are for an enhanced, not a reduced, fleet of freezer trawlers and the £1.06 million cost of refurbishment further confirms that rapid contraction was far from expected.[71]

Horobin and Tunstall observed fatalism or an omission to look ahead as characteristic of trawlermen. It is an attribute that can be extended to the distant-water industry as a whole.[72] Mumby-Croft and Barnard detect a short-termist or 'bonanza' mentality present from the earliest days of British distant-water trawling.[73] The established industry response to declining opportunities was to employ more efficient technology and/or find new grounds.[74] Self-regulation was untypical. Alan H., working for the WFA in the 1960s, remembers an industry with a stubborn confidence in the enduring bounty of the sea. 'There was no particular concern about stock depletion or stock loss,' he says, 'although the scientists were giving the signals. But there was so much fish in the sea and it was so productive, that these signals from the scientific community were largely ignored.'[75] The indicators may have been disregarded, but the reality of overfishing was not invisible to the industry. Alan believes that the Cod Wars forced a

69. C/DBTH/59/21. UK Fishing Fleet in the Mid-Seventies, pp.3–4.

70. C/DBTH/15/129. Re-location of Hull Fishing Fleet.

71. Sea Fish Industry Authority, Industrial Development Unit (SFIA) and University of Hull, Department of Geography, *An Analysis of the Hull Fish Industry in 1983*, August 1983, p.25.

72. Horobin, 'Community and Occupation', p.347; Tunstall, *The Fishermen*, p.134 and pp.171–175.

73. Mumby-Croft and Barnard, 'Antiquated Relationship?', pp.125–126.

74. Robinson, 'Steam Power', p.216.

75. Interview with Alan H.

re-evaluation that had long been on the horizon. 'I think really what happened at the end of the Cod Wars [*was*] that there was a rethink of what we were doing,' he reflects. 'We could no longer … go into the orchard and shake the apples down and just take away what we wanted. There was a limit to the number of orchards we could get apples from.'[76]

In adapting to new circumstances, the strengths and structures that had served Hull's fishery in the past, were poised to become weaknesses. Unlike Grimsby, with its mixed fleet of distant, near and middle water vessels, Hull ships were almost entirely for distant waters. In the shrinking seas of the 1970s, this restricted opportunities to diversify, particularly within the smaller regional fisheries of the British coast. A key problem was the size of the ships. From the mid-1960s Hull had amassed a modern fleet of freezer trawlers, the largest of which could freeze between 40 and 60 tons of catch per day.[77] These were highly efficient hunting machines with a voracious appetite for fish. From 1976, with increasing fishing restrictions and a growing anxiety over conservation, the instinct within the wider industry was to turn its attention to the smaller vessel. Alan H. explains, '[*Hull*] had big ships that … had a demanding need for a lot of fish to make a return. And I think really, the returns were getting less and less and these ships, to a large extent, were becoming white elephants. A much more profitable ship would be smaller and the sort of thing that you see operating off our coasts now.'[78] A 1979 *World Fishing* article confidently predicts the future elimination of the stern freezer trawler, just six years after those restructuring Hull's fish dock were foreseeing its expansion. The change of emphasis is ascribed to the appearance of fishing limits. The article asserts that whilst a small number of large vessels might continue to fish under reciprocal agreements overseas, the profitable future lay in a new breed of 'little ships', which would work efficiently in waters closer to home.[79]

In this context, Hull trawling firms not only faced the practical challenge of keeping large vessels profitable, they proceeded against a tide of industry opinion that was rapidly defining its cutting-edge fleet as oversized and anachronistic. In a conference paper delivered in October 1976, trawler owner Tom Boyd felt compelled to uphold Britain's continued need for

76. Interview with Alan H.

77. M. Thompson, *Hull and Grimsby Stern Trawling Fleet 1961–1988* (Beverley: Hutton Press, 1988).

78. Interview with Alan H.

79. *WF*, January 1979.

distant-water vessels.[80] Tom outlines the 38 per cent tonnage (45 per cent value) contribution made by the distant-water fleet to the total 1975 UK demersal landings. Without the robust distant-water trawlers, he argues, the coastal fleet would struggle to deliver an adequate or consistent supply to the shore-based fish-processing industry. Vessels with capacity to range across the North Atlantic could stockpile and maintain supply, unaffected by weather conditions that would render 80-foot boats incapacitated. Amidst the complex politics of fishing, Tom's plea was that the distant-water fleet should not be so readily dismissed. The 'big boats' needed a champion,[81] in the face of a wider fishing industry that, perhaps fearing the arrival of displaced fish-hungry giants in coastal waters, professed the 'little ship' as the future.

A further limiting factor for Hull's distant-water fleet was that its fishery was cod dependent, structured around the supply of cheap white fish to the mass market. Consumer demand for fish had steadily declined in the affluent post-war years, but by the 1970s, progress was being made with new cod-based frozen and convenience products, and although on a decrease, the fish and chip shop remained a British staple.[82] Just as British owners had not embraced the freezer trawler until driven by circumstances, as long as the market for cod continued, the industry saw little reason to alter its supply. In the late 1960s, Peter felt that the Hull industry was resistant to change. Peter was involved in an early experimental voyage to consider setting up a fishery in South Africa and was disappointed to find that the British industry were generally disinterested in the outcome. The reason, he feels, was that the fish was not cod.[83] In Hull, Peter also detected a business culture that was closed to the possibility of operating away from the city. 'I was left with the impression that the owners could not conceive that their vessels could not come back to Hull,' he explains. 'The thought of leaving your vessels in South Africa, servicing them, maintaining them and doing all that type of stuff and shipping the product back to Britain, didn't seem to prevail. It was, we've got to bring the fish back here to Hull.'[84] Alan J.,

80. *FNI*, December 1976.

81. HCA (HC) Ref: C/DBTH/20/74. BFF. Minutes of a Meeting of the Board of Directors held on 4 May 1978. An unnumbered agenda item contains papers relating to the Federation's adopted statement on the CFP. This includes a section entitled 'In Defence of the Big Boats'.

82. Reid, 'Trawler to Table', pp.161–163.

83. Interview with Peter.

84. Interview with Peter.

former accountant and director with the Marr Group of fishing companies, also felt that before the Cod Wars, the British distant-water fishery had been inward looking. He reports a strong industry focus on the home market, with certain species of fish rejected because they could not fetch a good price within the UK. 'They could've gone to land in Germany,' Alan reflects, 'but nobody really looked at what the other markets within Europe wanted. We were too inward looking by far and we'd almost reached the point where it was too late when we started selling fish into these [*other*] markets.'[85]

Before the 1976 catalyst, some small-scale new initiatives were being undertaken by the more progressive fishing companies. In the waters of South-West England, Hull firms were beginning to engage with the rapidly expanding mackerel fishery. Further afield, whilst the wider industry may have been indifferent to a South African fishery, in 1972, J. Marr & Son entered into a joint fishing venture with the Vesty Group in Cape Town.[86] A combination of unfortunate business and political circumstances led the venture to fail and such disappointments no doubt dampened any desire to exchange old practices for new. Trawling company Boyd Line had frequently been at the forefront of new technologies, including freezing at sea. However, Tom Boyd recalls times when his father's involvement with experiments had proved unproductive. 'It was all ahead of its time,' Tom explains, 'and it failed. So that helped convince [*my father*] … that one should just put your hand to the plough or the mast … and keep going and keep catching fish more efficiently.'[87] From this perspective, innovation was about timing. Good ideas at the wrong moment had brought bad results. Boyd Line, like other trawling firms, therefore concentrated upon trusted patterns of supply and demand. Whilst Northern seas remained open and home markets demanded cod, the distant-water industry continued as before.

There was a further hurdle on the home front. The British fishing industry was fragmented. Indeed, it could be better described as a series of industries.[88] Scottish, English and Northern Irish interests did not always coincide. The same was true of distant-water and inshore interests and the catcher and processing sectors. Even within the distant-water sector,

85. Interview with Alan J., March 2013.
86. M. Thompson, *The Marr Story, The History of a Fishing Family* (Beverley: J. Marr, 1995).
87. Interview with Tom Boyd.
88. *FN*, 25 May 1979.

tensions between the interests of local ports can be detected within the minutes of the BFF. In 1979, as the British distant-water sector campaigned for its place within the European CFP, outgoing Fisheries Minister John Silkin addressed the British fishery. The industry, he said, must understand 'that unless it remains united it will be utterly and completely destroyed'.[89] But the British fishing industry had little experience of cohesion. Looking back, Alan J. concludes, 'We were totally disorganised, and we came off very badly.' The Icelandic fishery was a major loss, but Alan feels that if the British industry had been well organised on entry to the EEC, there may have been a better outcome. 'But we came across as a disorganised and fragmented industry,' he reflects, 'and that's what did for us, in the end … When you can't agree amongst yourselves, what chance have you got with organised units like the French [*and*] the Germans?'[90]

Old Grounds with New Quotas

As Britain entered the post-Cod War era, the distant-water catching sector was forced to adapt to new ways of fishing. Traditional grounds were now redefined by territorial limits and operated under a sustained system of fishing quotas. Adjustment to these measures coincided with a period of unrelated commercial difficulties that augmented the adverse impact on the economics of trawling. Wage costs had been rising in line with other sectors of the national economy.[91] However, by far the biggest blow came from the oil crises of the 1970s. In 1973–74 and again in 1978–79, steep increases in the price of oil, resulting from global political instabilities, put severe strain upon the profit margins of industry worldwide. For distant-water trawling firms dependent upon oil for long voyages, as well as for the manufacture of trawling gear, the oil shocks sent costs soaring.[92] Coinciding with new fishing limits, it pushed the catching sector towards the brink of elimination. In 1972, the fuel bill for the whole UK fishing

89. *FN*, 25 May 1979.

90. Interview with Alan J.

91. Coopey and Woodward, 'British Economy', pp.4–5. Coopey and Woodward list the 1969 wage explosion as one of the factors influencing a worldwide downturn in business profitability in the 1970s. In *FN*, 30 May 1975, p.1, Associated Fisheries include wages on the list of rising costs that contributed to the organisation suffering half-year losses of £1.6 million in March 1975.

92. Robinson, *Trawling*, p.244.

industry had stood at £8 million. By 1974, it had risen to 27.5 million and was threatening to go beyond £33 million in 1976.[93] In 1975, the British Trawler Federation announced, 'the industry is bleeding to death. A collapse is imminent, and few seem to be noticing or caring.'[94] In Hull, memories of the oil shock stand secondary to the central issue of fishing limits. However, fish merchant Chris recognised the contribution of rising fuel bills to the problems of that time. The older and inefficient oil-burning trawlers, built when oil was affordable, were hit by spiralling costs and began to disappear. 'Cod War or not,' says Chris, 'you would have lost that part of the fleet anyway.'[95] The oil crisis stopped ships from going to sea, and Chris remembers seeing trawlers laid up in Albert Dock. Chris was witnessing the beginning of a fleet contraction that rapidly escalated after the final Cod War, but had its origins in earlier economic circumstances. By January 1976, it cost £250,000 per year to send a trawler to sea and the distant-water industry was announcing a £5.9 million loss. Oil and inflation were identified as the culprits.[96]

With hefty costs to meet, ships needed to be filled with fish. The constraint of the 50-mile limit had posed a challenge, but after 1976, conditions became increasingly problematic. In 1977, Norway, Russia, the Faroe Islands and Canada (Newfoundland) also extended their limits to 200 miles. Britain's membership of the Common Market meant that any rights to fish in third country water would be negotiated by the EEC and shared between member states. In the decade that followed, only small quotas were secured for the traditional grounds of the North-East and North-West Atlantic.[97] The European 'community pond' was itself controlled by setting a Total Allowable Catch (TAC) that was shared between member states as fishing quotas. However, politicised disputes over the size of the pond and the distribution of its fish delayed the production of a clear operating framework. It was not until 1983 that a consensus could be reached on how the TAC regulation should be distributed as national quotas.[98]

93. *FNI*, March 1976, p.21. It is not clear in the news article if these figures refer to the entire UK fishing industry or to the UK trawl industry. Nevertheless, the ratio of the increase indicates the scale of the problem.

94. *FN*, 30 May 1975, p.1.

95. Interview with Chris.

96. *FN*, 16 January 1976, p.1.

97. SFIA, *Hull Fish Industry in 1983*, p.20.

98. M. Leigh, *European Integration and the Common Fisheries Policy* (London: Croom Helm, 1983), pp.84–95.

For British trawling companies the small size of quota allocations presented a problem. In 1983, when the revised CFP finally emerged, Britain had secured the largest percentage share of the total catch in EEC waters. However, the allocation of opportunities for the British distant-water fleet to fish in third country waters was considered relatively poor.[99] A 1983 Sea Fish Industry Authority (SFIA) report into the position of the Hull fishery sees the 1983 CFP allowances as a clear official indication that the distant-water fleet was to be a mere '"token force" within the British fishing industry of the future'.[100] It defines the quotas given for distant waters as 'scarcely sufficient to warrant the journey unless combined with other quotas from nearby waters' and points out that the 1982 quota from Norway was barely enough to 'furnish each Hull freezer with at least two trips'.[101] With the entire North Atlantic subject to quotas, the owners of displaced vessels began to contemplate a shift to UK waters to claim non-quota species. In an article in *Fishing News International* in December 1976, trawler owner Tom Boyd asserted that:

> Lest one should imagine that [*British distant-water trawlers*] are, by their very size unacceptable, it should be noted that large French, Polish, Russian and Bulgarian vessels frequent these same 'UK pond' grounds.[102]

For those used to working the open seas, the quota system was frustrating, bewildering and inconsistent. More importantly, it made it difficult to fill big ships with a profitable catch. Tom Boyd recounts the problems faced by his own company in the shift towards new ways of fishing. In the years after the final Cod War, as vessels attempted to fish quotas of traditional species allocated from former grounds, the opportunities were gradually eroded. Tom says, 'our quotas came down and down. And ultimately, we finished up with just under 50 per cent of Britain's distant-water quotas, which we could catch with one and a half ships.'[103] The restricted and regulated pursuit of free-ranging shoals brought unforeseen challenges. As

99. SFIA, *Hull Fish Industry in 1983*, p.94.
100. SFIA, *Hull Fish Industry in 1983*, p.94. The UK is allocated 37.3 per cent of the catch in EC waters, but only 36.04 per cent of the total catch allocation, including catch in third country waters.
101. SFIA, *Hull Fish Industry in 1983*, p.20 and p.94.
102. *FNI*, December 1976, p.14.
103. Interview with Tom Boyd.

Tom explains, the movement of fish is not predictable, and the British fleet were accustomed to fishing wherever they chose across the North-East and western Atlantic. 'Once you've [*lost*] that ability to be flexible,' says Tom, 'and hunt the cod wherever you felt it was … it really was closing doors and making things extremely difficult.' The age-old divergence between scientists and fishermen came into the equation. Tom claims that for many years scientists would not believe the distances that fish could migrate. Cod, he declares, could migrate from Newfoundland to the North-East Atlantic. The scientists took a lot of convincing, because they found it difficult to prove. 'But our skippers knew,' says Tom 'because they'd followed it.'[104] However, limits now restricted where ships might follow, and pursuit of a quota could be frustrated by the migratory habits of different species.

It was not only fish that could prove inconsistent. Until 1983, the nascent quota system operated on temporary agreements. In 1978, trawlers were fishing on a 'makeshift' month-by-month Norwegian quota, which could be altered at any time, pending a longer-term agreement between Norway and the EEC.[105] In an increasingly global arena, quotas were vulnerable to international disputes unrelated to fishing. Tom recalls that negotiations with Russians for a much-needed cod quota came to an end when the EEC expressed disapproval of Russia's behaviour in its eastern provinces. A political stance wiped out the chance of Russian fish 'in a twinkling of an eye'.[106]

Back in Hull, whatever impacted upon the trawling companies affected the trawlermen who worked for them. In terms of trawling culture, skipper Patrick remembers the arrival of quotas as the end of the freedom of command. Patrick loved the thrill of the hunt, sailing up to Labrador and Greenland, using his own judgement to decide where to fish. Skippers were responsible to the trawling company, but once at sea, they were to a large extent their own masters. Quotas took away that liberty and with it the buzz of leading the chase. Quotas meant going where you were told and taking only what you had been allocated. Now, laments Patrick, it is the man behind the desk who is the skipper.[107]

For those still fishing, limits made it hard to make a living. Dave initially continued to work aboard the trawlers that continued to pursue

104. Interview with Tom Boyd.
105. *FN*, 24 March 1978, p.1.
106. Interview with Tom Boyd.
107. Personal communication (name changed).

white fish. But in the new order, it was not only trawling firms who found that the economics of fishing no longer added up. For crews too, ships were not catching enough fish to make a trip worthwhile. 'You felt your life'd ended,' Dave says. 'You think, where do you go from here? Do you carry on going to sea? Or do we just go back ashore, y' know, call it a day?'[108] To trawlermen fishing old grounds with new limits, it soon became clear that the traditional focus upon a small range of white fish types would not suffice. Distant-water companies needed to diversify and fish for mixed species. The mid-water trawl offered the means to be flexible. Tom Boyd describes a changing dynamic. 'We did everything we could to keep the ships operating,' he explains. 'We briefed every crew as to what was going on and they might go to the North-East Arctic one trip and then be going off herring fishing and going off mackerel fishing at the next trip.'[109] Flexible fishing took place with the same ships that had once ploughed the seas for cod. Boyd Line converted freezer vessels for mid-water trawling in the Arctic and enjoyed some success.[110] Other Hull companies also diversified, and trawlermen experienced in catching cod now turned their hand to new techniques. Skipper Bob describes his own experience:

> after the deep-sea bottom trawling finished, I went into pelagic fishing for mackerel and herring. Once the trawler owners realised that there wasn't gonna be any more – or very little – cod fishing, they had to turn to other types of fishing. And obviously they had to convert the vessels from bottom trawling to mid-water fishing. And as they changed that, so we had to change. It was either change or you were out, type of thing. And you just had to adapt to whatever was put in front of you ...[111]

Skippers unaccustomed to fishing in this way initially made mistakes that could result in ripped nets. It was trial and error, but the trawlermen proved to be versatile. As Bob makes clear, they had little choice.[112]

The challenges facing the distant-water sector were vast. Diversification came late in the day and amid major transitions within global fishing. In the struggle to adapt, there were victims and survivors. Ultimately, the

108. Interview with Dave.
109. Interview with Tom Boyd.
110. Interview with Tom Boyd.
111. Interview with Bob.
112. Interview with Bob.

seagoing sector would not be able to overcome the unfavourable circumstances that emerged in the void between the end of the final Cod War and the 1983 revision of the CFP. Limits, quotas and uncertainty combined to force Britain's cutting-edge distant-water fleet into terminal decline. To Tom Boyd, the lack of governmental support and a political inclination towards a small boat fishery further disadvantaged the British distant-water sector. In phrasing his recollection, Tom portrays the British government's response not merely as a failure to assist, but as an active hindrance. 'All the moves we took to keep the ships operating were regularly frustrated,' he concludes. 'We had hurdle, after hurdle, after hurdle placed in our way.'[113]

113. Interview with Tom Boyd.

Chapter Four

Living through the Aftermath

In the twenty-first century, territorial waters and fishing quotas are widely accepted as instrumental in the quest for sustainable seas. With the development of trawling technology and capacity, it is clear that unrestricted and unregulated access to North Atlantic fish could not have continued forever. But in 1976, in neighbourhoods threatened by the closure of open waters, the extent and severity of the forthcoming disruption had not been anticipated. The British distant-water industry sought to adapt, but significant losses could not be avoided. As the wake of global politics washed up on local shores, the impacts rippled through trawling port cities such as Hull. Hull's fishing district and culture had grown in response to an extreme maritime occupation. As the industry experienced upheaval, the trawling community faced the disintegration of their livelihood and lifeworld.

Ships to Scrap and Ships for Sale

Within community memory, the adverse conditions of the 1970s are closely associated with the scrapping or sale of the trawling fleet. As fishing opportunities decreased and in the absence of clear policy or aid to restructure, Hull's increasingly unprofitable distant-water vessels faced elimination. After 1976, Ken remembers the sight of trawlers laid up along the quay. 'The owners were getting rid of trawlers very quickly,' he recalls. 'You could see ships tied up, [*at*] one end of the dock … rows and rows of ships that were all tied up, not going to sea.'[1] If a message was needed, says Ken, that was it. As deeper decline followed, trawlers were towed away for disposal. Deckhand Joe was in his late teens at the time. He too recalls seeing ships not in service. He explains:

1. Interview with Ken.

> I'd say to somebody, what's the matter with them? [*and they would reply*], 'oh they're going for scrap' – and they'd be gone and there'd be a big hole in the dock. You could almost walk across the dock at one point [*across active berthed ships*], but then it just got smaller and smaller and to me … everything seemed to happen overnight.[2]

In Hull, the scrapping of trawlers often took place at Draper's Slipway, near Sammy's Point on the Humber Estuary close to the River Hull. Michael, sailing as mate in the growing absence of positions for skippers, witnessed the process. 'Each time you came home you'd see another trawler up there,' he recalls. 'It was really demoralising.'[3] The scrapping of ships was not a new occurrence, but something that had regularly taken place as trawling companies renewed and modernised their fleets. Thompson's inventory of Hull sidewinders working between 1946 and 1986 shows coal-fired ships being scrapped in the 1950s and 1960s with the advent of oil.[4] In the 1960s, the older oil-fired ships were scrapped to make way for the growing fleet of stern and freezer trawlers. The average life of a sidewinder was around twenty years.[5] In the decade before the 1970s, the sidewinder fleet had reduced by a half, from 140 vessels in 1961 to 73 in 1971, in the move towards modernisation (see Appendix 1, Table 3).[6]

However, as the 1970s progressed, factors other than age and technology began to have an impact. Overfishing in the North Atlantic put pressure on resources, limiting the catch. The 50-mile limit further reduced fishing opportunities and the added burden of the first oil crisis tipped the balance against the older elements of the fleet. As early as January 1975, the reduction of the Humber steamer fleet was being reported.[7] This was not

2. Interview with Joe, September 2013.
3. Interview with Michael.
4. M. Thompson, *Hull's Side-Fishing Trawling Fleet, 1946–86* (Beverley: Hutton Press, 1987).
5. Thompson's inventory shows the lifespan of most sidewinder trawlers to be in the vicinity of twenty years, with some exceptions. In calculating the fleet of the 1970s, the WFA assumes that vessels over 20 years old in 1975 would be scrapped (C/DBTH/59/21, UK Fishing Fleet in the Mid-Seventies, p.2). In the calculations for the move to Albert Dock, the size of the future fleet is calculated by the HFVOA on the basis of the withdrawal of a wet fish trawler after 24 years of service (C/DBTH/15/129, Re-location of Hull Fishing Fleet, p.6).
6. Appendix 1, Table 3, reproduced from Thompson, *Hull's Side-Fishing Fleet*, p.94.
7. *FN*, 28 January 1975, p.3.

yet a source of concern, for the old steamers had been living on borrowed time. But in the following years, oil problems in the wake of exclusion from North Atlantic waters posed further challenges to profitability. By 1977, oil-bunkering prices had risen to uneconomic levels and the last of the Humber steam trawlers were scrapped.[8]

The sidewinder fleet in Hull continued to shrink from 73 vessels in 1971, to 36 in 1976, to just five in 1981.[9] In business terms, these were the expendable vessels. Even before the industry entered crisis, the Albert Dock feasibility report had envisaged only ten wet fish trawlers remaining in the Hull fleet by 1987.[10] The future fleet predictions produced by the WFA in 1969 had assumed the removal of the oil-fired sidewinders by 1975, but with the diesel and diesel-electric side trawlers continuing into the later decade.[11] Yet Thompson's inventory shows that after 1976, it was these diesel-powered vessels that were being rapidly and perhaps unexpectedly sent to the scrapyard.[12] Nevertheless, the majority of the retiring diesels were already over twenty years old. The days of the sidewinder fleet had been numbered since the late 1960s. Limits, quotas and oil costs hastened rather than caused their departure.

In popular memory however, the story is different. Gill's photographic commentary on the Hessle Road fishing community during the late 1970s and early 1980s, contains the sub-heading *Ash and Dust*. Gill writes, 'This time of decay saw trawlers scrapped and houses smashed [*demolished*]. During the dual decline, fire and force brought an end to Hull's era as the greatest fishing port.'[13] The fires were those of Draper's yard, as ships were dismantled and contents burned. It is a plaintive contemporary image that over thirty years later remains emotive.

Skipper Ron remembers what he considered to be the premature end of the sidewinder fleet. He recalls:

> Oh, that was terrible … to see them, getting taken piece by piece. I mean a lot of them were still very good ships, y' know still had a lot of years of life in them. Some of them, the modern stern trawlers went off

8. *FN*, 10 June 1977, p.10.
9. Appendix 1, Table 3.
10. C/DBTH/15/129. Re-location of Hull Fishing Fleet, p.6.
11. C/DBTH/59/21. UK Fishing Fleet in the Mid-Seventies, pp.2–3.
12. Thompson, *Hull's Side-Trawling Fleet.*
13. A. Gill, *Hessle Road: A Photographer's View of Hull's Trawling Days* (Beverley: Hutton Press, 1987), p.93.

> to other places and were bought by other people, but the majority of the side trawlers were scrapped … and it was a really, really sad sight to see.[14]

In *The North Ships*, Piper describes visiting a Hull trawler pulled up onto the fish dock slipway for repair. He is disturbed by the motionless form of the typically life-filled vessel and concludes: 'She was nothing to do with me until they put her back in the water.'[15] A similar sentiment is present in memories of scrapping. Thomas conveys the hurt that he felt at the loss of the sidewinders, 'because I'd come up through the ranks on these ships. And to see 'em get … made into razor blades and stuff like that. It was destroying. Soul destroying.'[16] Like Piper, Thomas describes a troubling transition from the life-filled to the lifeless. Active ships that had shaped his life were reduced to the banality of razor blades, and it was devastating. Trawlers had been scrapped before. But this time it was different. This time ships were not to be replaced. In a 1962 photograph of Hull's first whole fish freezer trawler *Junella* making its initial appearance at St Andrew's Dock, over seventy men wait in anticipation around the lock pit.[17] In the past, the sadness of a fleet departure was tempered with the excitement of a new arrival. After 1976, this was not the case and in terms of popular memory, the scrapping of ships became a potent symbol of the end.

The scrapping of the old was followed by a more disquieting development. By the mid-1970s, economic circumstances were forcing the sale and conversion of Hull's advanced stern freezer fleet. 'That's when you knew, we won't gonna come back,' Thomas laments. 'It was always on a decline, and there was never gonna be an uphill any more … the slide was there and there was no stopping the slide.'[18] Since the Cod War negotiations in Oslo in spring 1976, the British distant-water fleet had struggled to make a profit from ever-decreasing fishing opportunities.[19] By 1977, a reduction in Norwegian quotas had forced the sale of the Hull-based Seafridge freezer fleet. They were sold to Norway.[20] Other transactions followed. In 1973,

14. Interview with Ron.
15. Piper, *North Ships*, p.26.
16. Interview with Thomas.
17. Thompson, *The Marr Story*, p.43.
18. Interview with Thomas.
19. *FN*, 18 June 1976, p.1.
20. *FN*, 10 June 1977, p.3.

Figure 6. The sidewinder *Joseph Conrad* being dismantled at Draper's scrapyard, 1980 (Copyright Alec Gill).

Henrikson & Co Ltd, founded in 1925, closed their operation with the disposal of their last two ships.[21] The remaining Hull companies persevered by engaging in new ventures, but the sale of trawlers nevertheless continued. In Appendix 1, Table 1 shows the steady leakage of advanced and modern vessels.[22]

There were some gains. In 1978, British United Trawlers transferred six of its freezers to Hull, leaving Grimsby without distant-water freezer vessels

21. Thompson, *Hull's Side-Fishing Fleet*, p.39.
22. M. Thompson, *Stern Trawling Fleet*, pp.128–129.

for the first time since 1964.[23] But the losses were significant. Earlier in 1978, the Boston Group fleet of ten seine netters, the only non-distant-water vessels in Hull, were sold to Consolidated Fisheries at Grimsby.[24] By March of the same year, distant-water trawler owners were facing the decision to sacrifice their wet fish fleets to keep the freezers active.[25] August saw only two of Hull's remaining 23 sidewinders at sea; it was described by *Fishing News* as 'a new peace-time low for the Humber port'.[26] By 1982, Britain's distant-water industry was sinking into deeper problems, with only the sale of assets holding back the debt.[27] In that year a total of 12 stern trawlers were diverted from the Hull fleet; the worst single year for such losses.[28] Amongst them were the *Arctic Galliard* and the *Arctic Buccaneer.* Owned by Boyd Line and launched as late as 1973, they were the largest trawlers in the British fleet, built with the highest freezing capacity of any ship in Europe at the time.[29] Their loss was a blow, as Tom Boyd explains:

> in addition to the pressure that we had from quotas, the banks were starting to get very frightened as to what was going to happen to the British fishing industry. And they withdrew support and there was a sort of wild scramble to get out, as one sees when banking fraternities feel they've got their money in the wrong area …
>
> So, we were under huge pressure from the banks to reduce our overdrafts and reduce any borrowing we had. And we were forced to sell the two top UK trawlers – the *Arctic Buccaneer* and *Galliard* … They were, you know, the pride of the UK fleet and we were forced to sell those to New Zealand, which was pretty depressing really.[30]

The cull of the British fleet continued into the late 1980s. By then Hull was unrecognisable as a distant-water fishing port. Policy delay and political stalling had condemned the British distant-water industry to a slow and convenient death. Financial support for the decommissioning of vessels arrived with the 1983 CFP. The aid was targeted to ships, not to

23. *FN*, 12 May 1978, p.2.
24. *FN*, 10 February 1978, p.1.
25. *FN*, 24 March 1978, p.1.
26. *FN*, 18 August 1978, p.2.
27. *WF*, March 1982, p.48.
28. Thompson, *Stern Trawling Fleet*, p.129.
29. *FN*, 26 March 1982, p.1.
30. Interview with Tom Boyd.

crews. It aimed to remove too many boats chasing too few fish. It did not compensate those whose livelihoods were caught up in the sacrifice. From an owner's perspective, Tom Boyd detected an astute political manoeuvre by the British government, with only a fraction of the fleet qualifying for support. 'The government of the day continually moved the goal posts,' he says. 'I suppose if you look at it in national expenditure terms, the industry was destroyed and removed without it costing the British taxpayer very much at all.'[31]

Laying Off and Staying On

The activity of Hull's fishing fleet had nourished a broad secondary and supply-based economy, bringing incomes into households throughout the city. As the ships went into decline, the impact was felt in the community. Anticipating change, some had left the industry before the crisis escalated, but many more stayed on. Ron and Victor had recently qualified as skippers and both hung on in the hope for a settlement with Iceland.[32] Shore-based radar engineer Philip had a mortgage and a young family. 'There wasn't a lot of work in my particular field,' he recalls. 'I did think of moving on and trying something else, many a time. But I stuck with it.'[33] Trawling and the fish dock offered a way of life that was familiar, and many sought to remain in the dwindling pool of opportunity. Deckhand Thomas continued fishing until it was no longer feasible, then in a complete career change, he joined the army. 'There was nothing else to do,' Thomas explains. 'What else could I do? … that's all I knew was fishing.'[34]

The desire to remain in a familiar occupation unites trawlermen and trawler owners. In the 1970s, Hull trawling firms included family enterprises handed down or managed from one generation to the next. Faced with a contracting industry, it was not only fishermen who knew nothing else. Trawler owner Andrew Marr explains that he remained in fishing because 'fishing is what we did, and my cousin and I relished the challenge'.[35] Tom Boyd too declares, 'We had people who depended on us. I enjoyed the

31. Interview with Tom Boyd.
32. Interviews with Ron, Victor and Pauline.
33. Interview with Philip, July 2013.
34. Interview with Thomas.
35. Notes from an interview with Andrew Marr, November 2012.

industry. Being an optimist, I hoped something could be salved.' On a personal note, however, Tom echoes Thomas, reflecting, 'You know, that's all I had done in my life, was manage fishing ships.'[36] The tenacity of Hull trawling firms would eventually lead to ventures that breathed a vestige of life into Hull's dying catching sector. But such efforts could not forestall the near-extinction of the fleet. As ships began to disappear, liquidation, closures and unemployment followed.

The full extent of employment within the Hull fishery is difficult to assess accurately. The industry engaged high levels of casual labour, making it hard to distinguish between actual employment opportunities and the numbers of employees that could potentially be contracted on an occasional basis. For example, the number of registered fishermen generally exceeded the number of berths on the trawlers. The industry created a range of secondary, ancillary and supply opportunities and it is not always clear to what extent these roles have featured in official industry figures. Finally, the spread of incomes generated by the fishery within the wider economy of the city (and on Hessle Road in particular) is difficult to calculate. The figures in Appendix 2, Table 1 are drawn together from various industry sources and provide a basic employment profile for the years between 1960 and 1993. The statistics for fishermen clearly show the impact of limits and quotas on the catching sector. Employment reduces steadily from 1960. The figures from 1973 are likely to reflect some of the earlier economic problems in the sector. However, from 1976, the totals decelerate markedly. By 1983, the numbers of fishermen have reduced by 74 per cent from 1976 levels, and by 1993 there is an 82 per cent reduction.

The figures for onshore employment are more open to interpretation. Results vary depending upon the information source, means of calculation and most specifically, which trades and functions are included in the inventory. In Appendix 2, caution is required in drawing precise comparisons, yet some general trends are clear. The onshore figures for 1960 are certainly underestimated, giving an unrealistic rise in employment in combined onshore trades between 1960 and 1972/73.[37] However, the downward trend in employment in the ancillary sector from 1972/73 is notable, along with the same accelerated decline between 1976 and 1993

36. Interview with Tom Boyd.

37. The SFIA data for 1960 draw upon Ministry of Agriculture and Fisheries figures, and exclude categories such as transportation, cold stores, independent shipbuilders, repairs and fleet supplies, which appear in some later inventories.

that is evident in seagoing jobs (a 58 per cent reduction by 1983, 73 per cent by 1993). Also apparent is the more moderate contraction in onshore fish-handling employment, when compared with the plunging figures of the catching sector. By 1993, the reduction is a less extreme 36 per cent from 1976 figures, revealing the resilience of the fish trades. As will be explained, by engaging with new suppliers, the fish trades were able to adapt to the changing situation. For the catching sector and ancillary trades this was not the case and as the challenges mounted, fishermen were amongst those hit the hardest.

Fishermen's Tales

The threat of unemployment within British distant-water trawling had escalated from the early 1970s. After 1976, job losses gathered pace in Hull as freezer trawlers joined the sidewinders in departing from the fleet. For fishermen working on casual contracts, their last days in the industry did not generally result from a formal notice of redundancy, but from fewer opportunities to sign back onto a ship or from their own decision to actively seek employment elsewhere. Ron worked for Newington Steam Trawling Co. one of the smaller companies on Hull's fish dock. He recalls that late in 1976, the company introduced a system of working four 21-day trips, followed by a one-trip spell ashore. It was a fair way of sharing the diminishing work, but it soon became problematic. Ron was sailing as a trawler mate and explains: 'You couldn't be guaranteed that you would just have the one trip off. [*Newingtons*] only had about five ships and if you had about seven or eight, nine, ten mates … sometimes there was quite a while before it came round to your turn again.' Crew members could find themselves spending long periods at home. 'At the end of the day.' says Ron, 'you've got to pay your mortgage, you've got to feed your family.'[38]

Under these circumstances, in 1977 Ron decided to quit. Ron was not alone and as departures escalated, Ron says that the industry imploded. 'People were leaving in the droves,' he recalls. 'They had nowhere to go, they had no ships.'[39] Whilst some left of their own accord, compelled by an obvious lack of a fishing future, for others departure was thrust upon them, as trawling companies finally folded. Deckhand Bill received a phone call ending his redeployment to Hull ships fishing for mackerel. 'They just

38. Interview with Ron.
39. Interview with Ron.

rung me up,' he says, 'and said that's it. Finished.'[40] Christine recalls how her husband Paul returned from a trip with the fishing company Thomas Hamling & Company, to discover that it was to be his last with the firm:

Christine:	He was onboard a ship … and he came home and he went back down dock to sign on the ship and all the firm was closed up … and that was a big worry, yeah. The firm'd just … it just closed up.
Jo (Researcher):	And there was no warning or notice given?
Christine:	No, nothing.[41]

Whilst those choosing to leave had the opportunity to plan their next move, those who found themselves suddenly out of work faced the dole and a search for new employment. Steve worked for the Department of Health and Social Security in Hull during the 1970s and 1980s. He explains that before the crises in distant-water trawling, fishermen were not regular visitors to his office. The fishing community was inclined to look after itself and Steve felt that for them, going to the dole office was a last resort. Traditionally, trawlermen had adopted a system known as backhanding, where money would be loaned to a friend or relative in difficulty, with the unexpressed expectation that the favour would be returned if necessary.[42] With so many leaving the industry or out of work, the usual systems of support were inadequate. After the Cod Wars, Steve observed a marked increase in benefit claims from the fishing industry. Claimants were not just from Hessle Road, but from many areas of Hull, where people worked in fishing-related jobs, and the workload of benefits staff rocketed.[43] In the close-knit world of fishing, whole families could find themselves out of work:

> It wasn't uncommon for a whole family of wage earners in a household to be thrown out of work at the same time. If dad's a trawlerman, if mother's working in a fish house and the eldest lad's a deckie learner

40. Interview with Bill.
41. Interview with Christine; likely referring to the St Jason, whose crew came home to Hull for Christmas in 1982, never to return due to the closure of Thomas Hamling & Co. Thompson, *Stern Trawling Fleet*, p.66.
42. Backhanding is raised in Steve's interview. It is also described in detail in Tunstall, *The Fishermen*, p.148.
43. Interview with Steve (name changed), May 2014.

> and the eldest daughter's working in a fish house, y' know, it doesn't take a great deal for a family who's been used to having quite a good standard of living to suddenly have nothing.'[44]

For fishermen facing unemployment there could be little financial cushioning to soften the blow. The need for benefits could be immediate, for the three-day millionaires were not inclined to save. Margaret, with a brother at sea and working herself in Hessle Road's Dee Street Club, observed at first hand that many fishermen had not planned for a rainy day. 'They wouldn't have thought, oh, wow, I'll have to put summat in bank,' she says. 'No, they never. They didn't.'[45]

The spectre of older redundant fishermen beached on the shores of Hessle Road is something remembered by younger narrators. Joe recalls men sitting in pubs and reminiscing, 'because they had nothing to do, especially the older ones who couldn't get work'.[46] Kevin was in his early twenties when the industry ran into difficulty, with time to move on to a new career. But Kevin too reflects, 'there was a lot o' kids [*in this sense, a local term for men*] who had done fishing all their life and that was it. They was finished.' He recalls men sitting on Hessle Road benches 'supping cider and wine, when their wives'd chucked them out, cos they couldn't cope with what was happening'. 'I know a lot o' kids went like that,' he adds.[47]

For those like Kevin, with years of work ahead of them, there were new opportunities that could be pursued. In his article on changing fishing communities in the 1970s, Lazenby questions the dominant post-Cod War image of redundancy, stagnation and state benefits.[48] Instead he highlights the mobility of fishermen and their ability to adapt. It is important not to minimise the impact of the crisis that emerged in British trawling ports, particularly in terms of unemployment. But in parallel, it is important to also acknowledge the competence with which trawlermen sought to adjust to new circumstances. Narrators for this book, from all sectors of the Hull fishery, exhibited resourcefulness and tenacity in seeking new roles. The demands of trawling-bred men

44. Interview with Steve.
45. Interview with Margaret.
46. Interview with Joe.
47. Interview with Kevin.
48. C. Lazenby, 'Change and Adaptation in European Fishing Communities, 1970–1997', in P. Holm and D. J. Starkey (eds), *Technological Change in the North Atlantic Fisheries*, Studia Atlantica: 3 (Esbjerg: NAFHA, 1999), p.195.

who were willing and able to work hard. It was a characteristic shared by other physically demanding trades on the fish dock. A strong work ethic could make former fishermen and fish dock workers attractive to new employers. This is illustrated in the narrative of John C., who was a trawlerman until 1958. He then came ashore and worked as a fish dock bobber, until made redundant in the aftermath of 1976. After two subsequent employment changes, John secured what he calls 'the best job I had in my life', working in the Reckitt & Coleman factory pool.[49] Supervisors were keen to have John in their team. 'I got sent from one department to another,' he says, 'and being in the fishing industry, you used to work like hell.' Tunstall has suggested that fishermen generally considered shore jobs as soft and lazy compared with the rigours of fishing.[50] Working outside the industry, John was surprised by incidents of procrastination, working to rule and long canteen breaks. Asked if he felt that the work ethic of the fish dock was advantageous to men leaving the industry to seek employment elsewhere, he answers, yes. 'Without a doubt,' he says, 'they was grafters. Really. I mean some of these young lads now, they'd die – die with their legs up, they really would.'[51]

The work ethic also saw men go to great lengths and distances to seek a new position. In the days after the Cod Wars, the networks of Hull's fishing community functioned as a surrogate job shop. Personal introductions, or 'putting a word in', had always been important in the fish dock employment market.[52] Word of mouth continued as a means of securing work in the circumstances after 1976. It was often a neighbour, a relative, a mate in the pub, who shared information about an opportunity. A new grapevine was emerging and this time it would extend all over the world. Closer to home, Margaret tells a story of her brother Harold's futile dash to Fleetwood in the hope of getting a ship. He and two friends followed a tip off picked up in Rayner's, the Hessle Road fishermen's pub. Margaret narrates:

> And off he trotted, they went, the three of 'em. And they said no, we don't want no Yorkies [*Yorkshire men*] here. So they jumped in a taxi

49. Interview with John Crimlis (John C.), July 2013.
50. Tunstall, *The Fishermen*, pp.170–171.
51. Interview with John C.
52. This is evident in many of the stories of how narrators got started in the Hull fishery. David also emphasises the role of local networks in recommending new recruits for his fish merchant business.

> [laughing] – no money to pay the taxi driver – and when they got to Harold's house, Harold emptied the gas meter in ten pence pieces and paid the taxi driver in a sock.[53]

The story resonates with other tales of tenacious pursuits in the fight to keep working. Some narrators gained jobs ashore, others at sea. Some stayed in fishing, whilst others transferred to new marine sectors. Some faced several months of unemployment before eventually finding a steady new career. Others relate a string of short-term and sometimes unreliable positions. But they did work, reinforcing the notion of adaptability, resilience, work ethic and perhaps a touch of fatalism. In many cases the new jobs could not match the financial rewards of trawling, but they offered a living. Deckhand John K. reveals the attitude that kept men working, explaining, 'Well, I always said, y' know, wherever I am, I've got two hands, I'll go and do any work.'[54] Ken offers his own conclusion: 'We were very versatile, fishermen, we could occupy almost any job and the training side … they'd find it very simple, y' know, cos they'd been through it all.'[55]

An Ancillary Story

Fishermen were just one fraction of the total workforce employed within the Hull fishery.[56] On land, an army of ancillary workers equipped ships for sea and despatch fish to the markets. Appendix 2, Table 1 provides an estimate of the ancillary workforce, although the calculation is not exhaustive. Through the ancillary sector, Hull's trawl industry spread deeper into the city economy, securing employment for thousands, from all walks of life and in all manner of trades. Ancillary roles included those that served the industry directly and exclusively: for example, the fish landers known as bobbers, employees at the ice factory and fish meal plant, ships' husbands or 'runners', clerks in the fish dock offices and trawler repairers and suppliers. However, there were also companies that served the fish dock on a partial basis: engineering, technical, repair or supply companies for which the fish dock was a major client, but which also served other marine

53. Interview with Margaret.
54. Interview with John K.
55. Interview with Ken.
56. Various ratios have been proposed for workers at sea to those on shore; the 1983 SFIA Report uses 1:6. SFIA, *Hull Fish Industry in 1983*, p.64.

Figure 7. St Andrew's Fish Dock looking west from the Lord Line building in 1966. The dock was home to an array of ancillary and secondary trades (Copyright Innes.co.uk).

and shore-based sectors. This included employers of white-collar workers, such as lawyers and accountants. Finally, there were offshoot activities that flourished in response to the needs of the dock and fleet: a myriad of tasks that may have evaded official statistics, but that nevertheless created work. There was the small boat owner who ran out into the Humber ferrying trawlermen who had missed a ship's departure and the man who collected and stored fishermen's belongings between trips. And there were the taxi drivers. Many trawlermen could not drive. Home for only short periods, there was little opportunity or inclination for driving lessons. In some households, women became the drivers, but for many trawlermen the taxi was their primary means of transport when on shore. Narrators cite the taxi as a staple of the trawling economy. A single taxi might stay at the disposal of a fisherman for his full time ashore. When Ron's father-in-law, also a fisherman, went to Beverley races, his taxi driver remained with him.

'[*The driver would*] turn up at the house,' says Ron, 'with his binoculars round his neck.'[57]

As the fleet went into decline, so too did the fortunes of this diverse and extensive economy. Just as in the catching sector, ancillary work began to reduce with the adverse conditions of the 1970s. By February 1976, for example, the trawler repair firm Humber St Andrew's Engineering was announcing its closure with the loss of 276 jobs. The reason for its demise, even before the end of the final Cod War, was the sharp reduction of fishing vessels operating out of Hull.[58] As the crisis deepened, the scale and pace of the losses increased. For the fish-landing gangs, the difficulties after 1976 came on top of gradual change and job reductions brought by the transition to mechanised landing. But compounding these was the decline in landing requirements that accompanied reduced fishing opportunities. Fewer ships, fewer landings, meant less need for landing gangs.

Other familiar companies began to close. Philip served as a shore-based marine electronics engineer with the Hull branch of Decca Marine Radar from the late 1960s. A large proportion of the company's work was on the Hull trawling fleet, the rest on commercial shipping. The fish dock was a big contract. Philip estimates that there was hardly a trawler in the fleet that did not have some Decca gear aboard. From the mid-1970s, as the remaining sidewinders disappeared, Philip's company absorbed the impact, as they removed equipment from ships due for scrapping. 'We had a depot full of radars, which had all been on a maintenance contract,' he recalls, 'autopilots, echo sounders, every piece of navigation equipment you could think of … we lost tons and tons of work, y' know, manpower and there were redundancies left, right and centre in most marine radio companies.'[59] The downturn continued into the 1980s. Decca made redundancies, but Philip managed to stay. His work, however, increasingly took him away from the fish dock and to commercial shipping contracts. By the mid-1980s the company staff had reduced by half. Around him, he says, joiners, fitters, boilermakers were made redundant from maintenance companies. As the Hull branch of Decca closed, Philip was redeployed to Grimsby. However, by the mid-1990s the Grimsby branch had also closed. In later years, gradual changes in science and technology would revise the way that maintenance contracts operated. But during his time with Decca, Philip

57. Interview with Ron.
58. *FN*, 20 February 1976.
59. Interview with Philip.

witnessed a contraction that resulted from the collapse of the trawling fleet. Philip was the last to go: '[*I*] put the lights off and put the key through the door,' he says. 'Yes, sad isn't it.'[60]

Philip's narrative reveals the extent and complexity of the trawling economy. It was not only maintenance contracts that provided Philip's company with work. Philip also fitted Decca systems for the brand-new trawlers that rolled off the slipways of the Humber shipyards. By the late 1970s, the changing fortune of the distant-water industry was taking a toll on shipbuilding too. In June 1977, *Fishing News* reported that in the Humber shipyards there was 'not a big trawler in sight'.[61] For yards already hit by the increase in large freezer vessels built away from the local region, uncertainty over the future of British fishing reduced orders to a trickle.[62] Whilst some were able to make up for losses with orders from alternative marine sectors, others could not manage the transition. The result was the closure of the local shipyards that had built the familiar names of the Hull trawling fleet. With them went orders for companies such as Decca. Beyond Hessle Road, a wider hinterland felt the gap left by the departure of its distant-water ships. Alongside engineering and radar companies, shops, pubs and taxi firms had all relied upon fishing to support their trade. Official industry figures may overlook how deeply rooted trawling had become within the formal and informal business of the city. Looking further afield, as the trawlers disappeared, the effects could be discerned way beyond the fish dock wall.

The Fish Trades: Less Cause for Concern

In 1976, the fish trades in Hull operated as a secondary economy to the primary function of fishing. Activity took place in proximity to the fish dock, where wet fish as a perishable foodstuff could be swiftly purchased, processed and distributed. Prior to 1976, the merchant and processing industry was experiencing change. Freezing at sea was altering fish landing, storage and sales practices. In parallel, fish was increasingly processed into packaged or frozen products for the retail market. The imperative for fish to be sold and prepared within the port of landing was beginning to decrease.

60. Interview with Philip.
61. *FN*, 10 June 1977.
62. *FN*, 10 June 1977.

Before the final Cod War, lingering traditions and practices maintained the impression that the fortunes of Hull's catching sector and its fish trades were inherently entwined. After 1976, this proved to be a fallacy. The fish trades suffered contraction, triggered by the removal of a regular supply landed by the Hull fleet. However, as the next chapter will show, they were able to adapt and carry on, albeit on a smaller scale, due to their ability to source fish from other suppliers. In the long term, the shift made Hull's fish trades vulnerable to global commercial and political influences. But in the years after 1976, as the catching sector faced crisis, for the fish trades there was less cause for concern.

Appendix 2 highlights the relative stability of the fish trades.[63] Chris and David both worked as fish merchants in their respective family businesses throughout the transitional years of the 1970s. Chris describes the Hull merchant businesses in the 1950s as largely locally owned. However, from the 1960s, there was a growing tendency for amalgamations, creating bigger businesses working across the major fishing ports.[64] David's business involved a small staff of filleters, who prepared and packaged fish for distribution, mainly to the fish and chip trade in Hull and other Yorkshire cities. In the 1970s, the Hull market was still dominated by the distant-water landings from Hull trawlers, with a mixture of wet fish and fish frozen at sea. There was also the North Sea market, where fish brought overland was sold in small quantities. David recalls that by the early 1970s, fish was beginning to enter the Hull market from Scotland.[65] In later years, this development was to become essential to the survival of the fish trades. But before 1976, it was Arctic cod landed by Hull vessels that defined the Hull fishery.

In the years after the final Cod War, Hull's fish trades faced severe disruption to their supply. Bound by limits and quotas the Hull trawling fleet could no longer meet the demands of its own long-held markets. To

63. Employment figures in the Hull fish trades show consistency across the research period. Palfreman's 1993 analysis asserts that employment in processing and the number of small primary processing firms in the Yorkshire and Humber region declined significantly in the 20 years prior to 1993. This is only marginally evident in the Hull figures in Appendix 2, particularly in comparison with other fishing sectors. Palfreman attributes the decline in processing to production line automation and the concentration of activities to meet new hygiene regulations, rather than to a drop in the importance of the industry within the region. D. A. Palfreman, *The Fishing Industry of Yorkshire and Humberside: A Regional Study.* Commissioned by the Yorkshire and Humberside Consortium, 1993, p.20.

64. Interview with Chris.

65. Interview with David.

survive, the fish trades turned to supplies from elsewhere. In the transition, there were casualties. Just as in other sectors of the fishery, David felt that many fish merchants had not taken steps to adapt their businesses in advance of fishing limits. The small size of many companies meant that it was not easy for them to instigate change.[66] As the volume of fish landed in the port reduced, there was a corresponding decline in work for the merchants. David explains, 'A lot [*of companies*] went down … there was a war of attrition and one by one they dropped out … if there was three warehouses full of merchants … year by year there got to be less and less. They couldn't cope.'[67] However, unlike elsewhere in the fishery, it was a contraction that progressed at a slower pace. The opportunity to source from elsewhere meant that fish merchants could successfully rescale and reorient their operations or withdraw gradually. Chris reflects that for merchants, 'it was a worry, but it wasn't a crisis, like it was for the catching side'.[68]

For smaller companies, the incentive and ability to survive often depended upon the willingness of the next generation to enter the business. The trade was made up of small family firms. 'If it's obviously becoming difficult,' explains David, 'then the new generation doesn't take it up and so you get all the people retiring.' David had a son who joined him and the family business continued for a further 40 years. 'But I mean many of them, their families drifted away to do different things.' David says. 'It wasn't the *in* thing. There wasn't a big future in it.'[69] Both David and Chris observed a similar trend amongst their employees, and as family-run firms scaled down, a smaller workforce was achieved through gradual wastage rather than redundancy. 'There'd been very little recruitment into the industry, I would say since the 70s,' Chris recalls. 'I'm still in the industry and there are very few of us left, and they're all my age. There are few youngsters … whereas in the [*19*]60s there was every age, from teenagers to people in their 60s.'[70] Those that did continue proved to be adaptable. David recalls that 'in the darkness of the night, you'd sometimes wake up and think will we be able to go on'.[71] But both men were confident that their business had a future. Hull trawlers may have gone, but the fish trades could turn

66. Interview with David.
67. Interview with David.
68. Interview with Chris.
69. Interview with David.
70. Interview with Chris.
71. Interview with David.

elsewhere. They could adapt and survive, because as Chris observes, 'There's always somebody, somewhere wanting to send fish.'[72]

Much larger food companies dealing in frozen and oven-ready products also persevered. Firms such as Birds Eye, Findus, Ross and Marr Frozen Foods, retained their workforce, yet they faced difficulties with supply, which at times brought the threat of closure.[73] Fish processor Ivy recalls that from the mid-1970s there were fewer opportunities to work. She says, 'It was sort of like a hit and miss. You either was there because there was the fish or you had to have a week off because there wasn't any.'[74] Ivy reports instances where work was so sporadic that workers decided to find jobs elsewhere. However, in the years after 1976, like David and Chris, Ivy describes a contraction, not a collapse. Factory work continued, there was just 'less of it'.[75] Billy, from his fishmonger's shop on Hessle Road took a philosophical view. To Billy, the fish trades had a way of carrying on. 'They always seem to do in Hull … even now. There's not much of it left. No smoke house, no fish dock. [*But*] still a few firms doing a bit.'[76]

In the 1970s, Hull suffered a jolt that stemmed not from the troubled national economy, but from the changing politics of fishing. The disruption brought social and economic consequences. As the fish trades struggled on, the catching sector and its ancillaries faced a terminal decline. Trawling company director Alan J. captures the sense of loss and anxiety that pervaded the city. 'I think it hit quite a few people,' he says. 'If you'd have asked people in the 60s, how long the fishing industry'd been there, they'd have said forever. And of course, it went steadily downhill.'[77] The crisis brought a sense of precarity. Even those outside the industry perhaps felt

72. Interview with Chris.

73. A newsletter to staff at the Birds Eye factory in 1978 appeals for them to avoid strike action at a time when the factory was already facing problems outside their control, such as fish supplies. Source: N. Triplow, *Pattie Slappers: Stories from the Golden Age of Hull's Food Processing Industry* (Immingham: Community Heritage Publishing, 2013), p.78. Gerry Raines, former manager with Marr Frozen Foods also comments upon the difficulties of fish supply coming into Hull in the 1980s. Source: Triplow, *Pattie Slappers*, p.73. In 1979, Findus in Hull was threatened with closure due to supply difficulties. Colin Birch, Findus technical director, blamed a deterioration in the overall UK fishing situation resulting from the uncertainties surrounding the CFP. Source: *WF*, June 1979, p.95.

74. Interview with Ivy.

75. Interview with Ivy.

76. Interview with Billy, April 2013.

77. Interview with Alan J.

less secure, having witnessed how quickly fortunes could change. As Alan reflects, 'a lot of people always thought, y' know, they got in these bigger companies and [*they thought*] they were there for life and they weren't.'[78] The impacts of the Cod Wars revealed the fragility of former assumptions. In the aftermath came the need to engineer new lives.

78. Interview with Alan J.

Chapter Five

New Horizons

To counter contraction and loss, there was a pressing need for Britain's distant-water industry to revise their operations and seek new opportunities. Prior to the final Cod War, the sector had embraced emerging technologies. However, the focus had remained conservative, fixed upon traditional grounds, species and markets. In Hull, it took the challenges of the 1970s to drive the city's once coherent fishery in new directions. Until that point, the industry had projected measured progress and permanence. Yet 140 years earlier, trawling from Hull had been founded as market pressures on the south coast sent fishermen north. Then, exploratory journeys had led to the discovery of the Silver Pits and subsequent expansion into the North Atlantic. In the 1970s, descendants of those early entrepreneurs remained amongst the leaders of Hull trawling companies, as once again, critical pressures led them towards unfamiliar practices, alternative fisheries and new maritime sectors.

Immediately after the Cod Wars, Hull trawling firms remained committed to the business of fishing. In the 1970s, in his role as group accountant for the Marr company, Alan J. faced the difficult task of making multiple redundancies across the trawling ports of Hull and Fleetwood.[1] He observed the dependency of fishing communities upon the trawling companies, recalling that 'people clung on to the hope that … we could do something'.[2] Although slow to anticipate the crisis, once faced with its consequences, trawling firms responded with tenacity and energy in efforts to reorient their business and keep the industry afloat. Embarking on new ventures, Hull owners generally retained their regular crews, who were known and trusted, were familiar with the ships and turned out to be remarkably adaptable.

1. Interview with Alan J.
2. Interview with Alan J.

There were unexpected challenges. During the 1970s, the BFF had maintained its commitment to recruitment and training, even amid the shedding of jobs and vessels.[3] However, as the fleet scaled down, there came the danger of not having enough crew for the remaining trawlers. The lack or irregularity of work had driven fishermen to alternative employment and the absence of an EEC fishing policy made it impossible to predict stable long-term requirements.[4] By 1982, the Registrar for the Humber Freezer Trawler Owners Company (HFTOC) was reporting a serious situation: 'Good deckhands are not willing to sit around waiting for a fishing job,' he observed, 'they are seeking and obtaining employment in other sea-going occupations.'[5] The casual basis of employment made it difficult to keep track of who remained as a potential employee. The Registrar's list of available crew for February 1982 records 11 men who were 'unofficially reported to me as left'.[6] In that year, HFTOC set up a centralised register to proactively track the availability of men.[7] For firms accustomed to the pick of good crew, the roles had been reversed. The problem was not just one of numbers, but of maintaining skills and experience. Tom Boyd recalls his concern that there would not be enough skilled bosuns, mates and deckhands for the time when he hoped to piece his company back together. Crew had left. 'We had to trawl around,' he says, 'and get people back.'[8]

3. In March 1977, the Federation were asked if applications from port authorities to fund workforce recruitment and training should be supported given present circumstances within the industry. The Federation's view was that 'it would be a folly to reduce our training and recruitment commitments', which were necessary 'to produce the right type of crew member for the future'. HCA (HC) Ref: C/DBTH/20/72. BFF. Item 77/55 in Minutes of a Meeting of the Board of Directors held on 3 March 1977.
4. HCA (HC) Ref: C/DBTH/15/115. HFTOC. Minutes of a Meeting of Trawler Managers and Shipping Masters held on 19 November 1981.
5. HCA (HC) Ref: C/DBTH/15/115. HFTOC. Letter from Registrar to all Directors dated 12 February 1982.
6. HCA (HC) Ref: C/DBTH/15/115. HFTOC. Letter from Registrar, 12 February 1982.
7. HCA (HC) Ref: C/DBTH/15/115. HFTOC. Minutes of a Meeting of Shipping Masters and Trawler Managers held on Thursday 26 November 1981.
8. Interview with Tom Boyd.

The Search for Alternatives

Home Waters

As Tom had forecast in 1976,[9] the failing battle to make a living from distant-water white fish quotas soon brought the British distant-water fleet into home grounds. With low demersal quotas, trawling companies, aided by the versatility of the mid-water trawl, turned their attention to other species and to pelagic fishing in particular. Their focus fell on non-quota species, which initially could be freely fished. In parallel, companies engaged in experimental fishing, trawling for hitherto unexploited species, including those that existed in the deep waters off the British continental shelf. For Hull firms, UK-based ventures offered new possibilities for former Arctic ships and crews, whilst for some Hull trawlermen, an emphasis on home waters meant the chance to work for alternative employers in the diverse coastal fisheries of the British Isles.

'Mackerelling'

The single most important opportunity for the Hull trawling fleet arose from the south-west shores of England. From the mid-1960s large overwintering shoals of mackerel had appeared off the Cornish coast. Between 1970 and 1975, mackerel fishing developed as a 'boom industry', with landings in England and Wales increasing from 3,000 to 30,000 tons.[10] Since the mid-1960s handlining had been the predominant UK catching method, but from the 1970s, trawlers were also taking a share. Scottish trawlers and seine netters, displaced by heavy herring fishing restrictions in the North Sea, were becoming regular visitors to Cornwall, increasingly staying for a complete mackerel season. In 1975, for the first time, the combined UK trawler and purse seiner catch exceeded that of the handliners.[11] The increase in the UK landings was exceeded by the huge amounts of mackerel taken by the foreign fishing effort. In 1970, the total mackerel catch by all nations reporting to the International Council for the Exploration of the Sea for the English Channel and Celtic Sea was 60,000 tons. By 1975 this

9. *FNI*, December 1976.

10. HCA (HC) Ref: C/DBTH/70/28. Mackerel Research in the South-West. Ministry of Agriculture, Fisheries and Food: Directorate of Fisheries Research. Laboratory Leaflet No. 32. Lowestoft, 1976, p.2.

11. HCA (HC) Ref: C/DBTH/70/28. Mackerel Research, pp.2–3.

had exceeded 300,000 tons, with vessels from Eastern bloc countries and the USSR becoming the major new arrivals on the scene.[12]

As questions were raised about the long-term impact of this incursion, the developing mackerel bonanza was watched with interest from the Humber. Trawling firms contending with restrictions, quotas and perceived unfair competition had begun to consider new options. In October 1975, Boyd Line sent the UK giants *Arctic Galliard* and *Arctic Buccaneer* to join the mackerel fishers.[13] Other Hull firms followed. Northern ships were heading south and the move was greeted with dismay by the Cornish inshore industry. In August 1976, *Fishing News* reflected: 'Although the Cornish fisherman in his 25ft line boat might have thought the Cod War with Iceland or the row over North East Atlantic Fisheries Commission (NEAFC) quotas for herring had little to do with him, he is now finding out otherwise.'[14] From the Humber perspective, however, there was logic, even justice, in relocating the distant-water fleets. The move to the south-west occurred in the face of shrinking fishing opportunities in the North-East Atlantic. Amid these circumstances, and despite the lower market value, there developed a sense that access to mackerel could compensate for the losses the British freezer fleet had suffered elsewhere.

Initially, mackerel was fished without restriction. However, the mounting pressure on stocks led to controls, beginning with the Mackerel Protection Order in 1977, and progressing to restrictive licensing from 1980.[15] In the beginning, Hull firms extracted mackerel for a reduction fishery. Fish was transported back to the city by rail to the Hull Fish Meal and Oil Company and plans were underway to export to the Danish reduction industry.[16] The arrival of the Mackerel Protection Order put restrictions on this practice. The restrictions, however, coincided with a significant new opening. In 1977, the declaration of a UK EEZ saw fleets from Japan and Eastern bloc states, with no reciprocal fishing opportunities to offer, excluded from the newly created 200-mile European pond.[17] Leigh asserts

12. HCA (HC) Ref: C/DBTH/70/28. Mackerel Research.
13. *FN*, 31 October 1975, p.3.
14. *FN*, 13 August 1976, p.9.
15. Sea Fish Industry Authority (SFIA), Industrial Development Unit, *The Importance of Klondyking to the UK Pelagic Fishery*. Internal Report No. 1272. May 1986, p.8.
16. *FN*, 16 September 1977, p.1 and 23 September 1977, p.3.
17. Leigh, *European Integration*, p.91.

that this did more than any other measure to increase possibilities for EEC fishermen.[18] The move changed the operations of the UK fleet trawling for mackerel in the south-west. By August 1977, approval had been given for a 'red fleet' of colossal Eastern European factory ships to be based on the Cornish mackerel grounds.[19] The ships would not fish, but would receive British-caught mackerel, transferred in nets or pumped aboard, to feed their enormous processing capacity. Known as klondyking, this method of operating for export became the cornerstone of the UK mackerel trawl fishery in the 1970s.

Alongside klondyking, Hull companies were also engaging in the direct export of mackerel frozen at sea to African markets. Andrew Marr of the Marr Group recalls the strong demand for frozen fish in Africa, where fish had traditionally been consumed salted or dried. Andrew's negotiations in Nigeria led to the development of J. Marr (Seafoods) Ltd as the fish-trading arm to the company's other interests.[20] In the south-west, the progression from reduction fishery to fishing for human consumption is a watershed in Hull's post-Cod War development. For the first time, Hull ships fishing for mackerel were pursuing large catches intended almost exclusively for an export market. Hull-caught mackerel was going to Eastern Europe, West Africa and Japan. As the surviving freezer fleet fished between pelagic and demersal species, Hull ships returned less frequently to their home port with a catch bound for UK markets.

For Hull trawlermen wanting to continue in fishing, mackerel offered an important lifeline. It was a new way of working, and even for those well-practised in Arctic trawling there were lessons to learn. Pelagic fish shoal in high numbers. Alan J. recalls that initially mackerel was caught in excess of the trawler's freezing capacity. Skippers were instructed to take just a fraction of the shoal: no more than could be processed and frozen before the oily fish became rancid. But for men accustomed to pursuing the big catch, old habits were hard to break. 'If they were on a fish shop as they used to call it,' says Alan, 'they'd catch as much as they can.'[21] Cultural conditioning aside, grabbing a small corner of a big shoal was not easy. Bob, a trawler skipper, remembers the appearance on the fish sounder of a black ball indicating mackerel. 'There could be hundreds and hundreds and

18. Leigh, *European Integration*, p.91.
19. *FN*, 5 August 1977, p.1.
20. Interview with Andrew Marr.
21. Interview with Alan J.

hundreds of tons in that black ball,' he explains. It was a case of clipping the edge. 'You couldn't actually tow your net through that ball,' Bob recalls. 'You'd finish up with no trawl.'[22] But although the fish were prolific, deckhand Bill missed the variety of Arctic fishing. Arctic cod demanded a hunt and if you caught it, you felt lucky. To Bill, mackerel was perhaps too predictable. 'It was [*just*] there,' he recalls. 'Once you were on top of the mark, you knew that trawl was full.'[23]

As an oily fish that deteriorates quickly, mackerel was transhipped or frozen whole very soon after being caught. Dave also remembers the sheer size of the hauls, saying, 'I'd never seen a bag of fish like it.' As a trawler engineer his concern was how to process it. Freezing pelagic species in a temperate climate was different to handling cod caught in cold seas. There was a constant battle against heat, generated by the oily fish and aided by the ship's environment. It was a learning curve and Dave recalls onboard disagreements over the correct approach.[24]

The introduction of klondyking helped to overcome the limitations posed by a trawler's freezing capacity. Bob describes a production line of Hull trawlers transferring mackerel to Russian and Polish vessels anchored out in the bay. 'The company use to say, go to so-and-so ship, he's your next ship. And then you went alongside and he just took all the fish off you and then you left that one, went and caught some more fish … and that's how we carried on.'[25] By contrast, fish that was frozen at sea was run into local ports. In South Wales, Milford Haven became an important transhipment port, where fish was reloaded into refrigerated vessels, much of it bound for Nigeria. By the late 1970s, all the Hull trawling companies were working on the south-west mackerel grounds. The HFVOA rented accommodation and facilities from the Milford Docks Company. The result was a Hull colony in the Welsh port. The arrival and departure of Hull ships was no longer confined to Hull's own fish dock.[26] Trawlers generally stayed in the south-west for the entire mackerel season and crews left and returned to Hull by coach or by taxi. With ships away, there was even less work in

22. Interview with Bob.
23. Interview with Bill.
24. Interview with Dave.
25. Interview with Bob.
26. HCA (HC) Ref: C/DBTH/15. HFVOA. The minutes of the governing bodies record increasing activity and involvement with premises and facilities at Milford Haven from mid-1977.

Figure 8. 'Once you were on top of the mark, you knew that trawl was full.' Mackerel on the deck of the Marr freezer trawler *Cordella*, 1980 (Copyright Alec Gill).

Hull for the already diminished ancillary trades. However, engineers too travelled to service equipment aboard the Hull fleet, working alongside local Welsh companies. Ben remembers seeing familiar companies and faces in the Welsh port – fitters, electricians and shipwrights. 'It was like Little Hull down there sometimes,' he says, laughing.[27]

The Hull migration was not welcomed by all. Tensions between the expanding trawling fleets and the local inshore fishermen gathered pace over the decade. The BFF handled various complaints; freezer vessels processing at night with lights blazing; large ships trawling through fleets of small mackerel-jigging boats.[28] With the mackerel fishery crucial to their attempts to repurpose, the Federation strove to avoid confrontation that could see them banished from the six-mile limit and spell 'a further nail in

27. Interview with Ben (name changed), April 2013.

28. HCA (HC) Ref: C/DBTH/20/73. BFF. Memorandum, Agenda Item 6 for Meeting of the Board of Directors to be held on 1 December 1977.

the coffin of the trawling industry'.[29] Nevertheless, the colossal rise in catch levels inevitably brought the threat of overfishing. Mackerel quotas were introduced from 1977. However, ongoing miscalculations of fishing effort, vague and optimistic stock assessments and the ambiguous status of TAC recommendations in the absence of an agreed fishing policy resulted in poor regulation, confused guidance, the early halting of fishing seasons and the reduction of quotas.[30] In 1979, to protect the seriously depleted stock, freezer trawlers and seine netters were excluded from a government-imposed 'Cornish Mackerel Box', and the fleet's attention was redirected to the mackerel stocks in the north-west of Ireland and the Outer Hebrides.[31] In 1982, the government decided to withhold licences to the Eastern bloc factory ships that had been visiting the Cornish mackerel grounds since 1977.[32] The need for regulation was clear, but the chaotic implementation once more saw Hull ships unable to make a profit. As fleets turned once again to the north, the Hull ships missed out on quotas, which were directed to the Scottish fishery. Company owner Tom Boyd witnessed his own allocation come 'down and down,' until, he says, 'we just didn't have sufficient quotas to keep the ships operating'.[33]

For Hull trawlermen, falling mackerel quotas once again meant reduced sea-time and low earnings. When Dave could anticipate just three to four weeks of work per year, he finally decided to come ashore. 'That's when I gave it up,' he says. 'I thought, this is just not worth it. There's no work. Y' know, you weren't allowed to work.'[34] But for almost a decade after the final Cod War, the south-west mackerel fishery helped to keep Hull trawlermen at sea and provided employment for the depleted ancillary trades. It was a different experience to that of the Arctic. In Hull, it developed its own parlance. The term 'mackerelling' came to signify novel techniques, new markets, warmer seas and overland journeys to southern ports. It is a term that when spoken conveys something different to 'fishing'. Fishing meant departing from Hull for distant icy waters and for cod.

29. C/DBTH/20/73. BFF. Memorandum, 1 December 1977.
30. For example, see *FN*, 23 February 1979, p.1; 14 September 1979, p.3; 11 January 1980, p.1; *WF*, December 1982, pp.18–19.
31. SFIA, *Klondyking*.
32. *WF*, December 1982, p.19.
33. Interview with Tom Boyd.
34. Interview with Dave.

Red Herrings and Monsters from the Deep

Mackerel was not the only species considered by the UK distant-water industry as an alternative to cod. From the mid-1970s, with pressure on cod stocks and fishing limits looming, the British government had funded research into potential new fishing operations within its home waters. Their approach to developing new fisheries was more than a question of extracting the resource. Attention was given to cost, to fish handling and processing, and to available or potential markets. Consumer preferences were considered, which in Britain meant working against an extremely conservative fish-eating tradition. Nevertheless, the process began with the identification of available fish. Here the WFA and other research agencies worked in consultation with the BFF, and the resulting exploratory journeys often took place in the vessels of interested Hull trawling firms.

By 1980, alongside mackerel, potential investment opportunities had been identified for horse mackerel, sprats, pilchards and an array of shellfish.[35] The government had rejected research into the pursuit of Antarctic krill.[36] Instead, they pinned their biggest hopes on blue whiting, which could be exploited much closer to home. Blue whiting, a pelagic fish of the cod family, existed in significant quantities in very deep water off the continental shelf to the west of the British Isles.[37] Prior to 1974, there had been little fishing effort on the stock,[38] and it was hoped that markets could be developed, not only for fish meal but also for human consumption. In 1974 and 1975, the Hull vessels *St. Benedict* and *Arctic Princess* were engaged in blue whiting exploratory voyages.[39] Yet whilst some firms were keen to get involved, others were less enthusiastic. In March 1976, amongst members of the BFF were those who felt that defining fishing limits should be their priority, and there was a consensus that the industry should 'keep relatively quiet on blue whiting fishing until it was possible to regard it as a

35. HCA (HC) Ref: C/DBTH/20/78. BFF. Agenda Item for a Meeting of the Board of Directors to be held on 7 February 1980. Programme for Improving the Processing and Marketing of Fish and Fish Products in England and Wales under EEC Regulations 355/77, points 68–93.

36. *FNI*, August 1976, p.21.

37. HCA (HC) Ref: C/DBTH/70/32. Blue Whiting. Ministry of Agriculture, Fisheries and Food: Directorate of Fisheries Research. Laboratory Leaflet No. 45. Lowestoft, 1979, pp.1–2.

38. C/DBTH/70/32. Blue Whiting, p.2.

39. *FNI*, September 1976, p.97.

recognised food fish'.[40] Nevertheless, other firms retained their involvement in the experiments. By the late 1970s, progress had been made in the battle to process these small bony fish into marketable products, and consumer trials had received promising feedback from the reticent, cod-loving British housewife.[41]

As other nations, Norway in particular, showed an interest in blue whiting for fish meal, the British trawling industry needed to establish an historic preference for the species, to gain a foothold against future competition.[42] However in 1978, UK catchers lost considerable sums in an experimental commercial blue whiting season.[43] As the fish disappeared within the non-accessible Faroese limits, Hull's Boyd Line lost £250,000 and thenceforth withdrew from the fishery.[44] In 1979 and 1980, significant blue whiting landings by Hull trawling firms Marr and Hamling proved that with the right conditions the fish could be caught.[45] However, the catch sold as pet food and fish meal. The outcome was a far cry from the anticipated new consumer fishery, and the distant-water industry's interest was waning.

Alongside non-traditional but recognisable species such as blue whiting, more extreme experimentation was also taking place in the very deep waters off the continental shelf. Here fish with strange names and unfamiliar profiles were pulled from the depths in the hope that a viable fishery would be discovered. But if mackerel and blue whiting presented a challenge to the British consumer, then the likes of scabbard and monkfish were unlikely to make it to the table. Moreover, the complexities of catching and processing these fish further conspired to keep such species off the menu. The recollections of those involved in experimental fishing in the 1970s illustrate the reasons why their endeavours were unable to reinvigorate the industry. Alan H. worked on several WFA investigations. The species living 'down the cliff face' of the continental shelf were, he says, 'edible, but not particularly marketable'. Orange roughy, successful in New Zealand, could

40. HCA (HC) Ref: C/DBTH/20/69. BFF. Item 76/38 in Minutes of a Meeting of the Board of Directors on 4 March 1976.

41. *FNI*, January 1977, pp.63–67.

42. HCA (HC) Ref: C/DBTH/20/74. BFF. Item 78/02 for Minutes of a Meeting of the Board of Directors on 12 January 1978.

43. HCA (HC) Ref: C/DBTH/20/75. BFF. Memorandum. Agenda Item 8 for Meeting of the Board of Directors on 5 October 1978, p.3.

44. *FN*, 6 April 1979, p.1.

45. *FN*, 18 May 1979, p.2 and 25 April 1980, p.1.

be found in deep water off North-West Britain. But as Alan recalls, it was a difficult process. 'You were really expecting people to be fishing at depths of 2,000 meters or 3,000 meters,' he says, 'and it's a long way to put a trawl down and bring it up empty.'[46]

It wasn't just trawling companies that needed to be adventurous. Fish merchants David and Chris reflect upon the challenges of enticing the British consumer. To Chris, the experiments were destined to fail. 'No pun intended,' he says, '[*but*] it's a red herring. Because, tell me this, have you ever eaten any unusual fish? ... when it comes to eating fish, we [*the British*] are very conservative.'[47] David lists some of the unusual names amongst the fish being investigated; scabbard, monkfish, rabbit fish, rat-tail. 'You couldn't produce that into a pack to go into the supermarket,' he explains, 'because simply nobody would buy it.'[48] The appearance of these unusual fish caused one fish sales representative to define them as 'monsters from the deep'.[49] Coupled with unappetising names, strong flavours and limited usable flesh, there was little to endear such species either to the British consumer or to the industry.

Blue whiting, the strongest contender for the UK consumer market, also proved problematic. Alan J. recalls the Marr Company's involvement. 'We caught a couple of hundred tons and Birds Eye and Findus and one or two other companies played around with it,' he says. 'Birds Eye and Findus particularly thought they could use it to turn into fish fingers, but it was very, very small fish.'[50] The small fish had to be processed by hand, which added to labour costs. Thomas, as deckhand, remembers the unpopularity of the blue whiting with those processing the fish at sea. 'I think the biggest fish was about three inches long,' he recalls. 'You imagine trying to gut a ton – one ton – of these fish at three inches long. It took us about a day and half, two days and that was everybody.'[51] Thomas narrates that on

46. Interview with Alan H.
47. Interview with Chris.
48. Interview with David.
49. Interview with Tom Boyd.
50. Interview with Alan J. In 1979, the Findus annex in Wassand Street, Hessle Road was the centre for a major blue whiting-processing experiment, putting Findus and Hull at the forefront of the potential new fishery. *WF*, June 1979, p.95.
51. Interview with Thomas. The Torry Research Station publication *Handling and Processing Blue Whiting* suggests that the average size of blue whiting from the British north-west coast is between 25 and 36 centimetres. See www.fao.org/wairdocs/tan/x5952e/x5952e01.htm (accessed 14 July 2014).

one experimental journey the crew's dislike of gutting blue whiting led to dispute. Back ashore its small size helped to make blue whiting unviable as a commercial foodstuff. As Alan J. explains, the labour cost was 'too great a proportion to the final product and it rapidly became obvious that there was nothing we could do with the blue whiting as far as human consumption was concerned'. It was considered for fish meal, but in the end that too was felt to be unworkable.[52]

During the 1970s, Hull companies, ships and crews played an important role in the search for new UK fisheries. Ultimately their participation did not result in the hoped-for long-term opportunities. It is tempting to speculate that had the investigations taken place a decade earlier, alongside a functioning cod fishery, then the challenges may have been more easily accommodated. In the late 1970s however, losses from experimental fishing came on top of other obstacles facing the industry, and its continuation was untenable. Yet it was precisely these circumstances that had spurred the new thinking. In the absence of crisis, it is questionable if an attempt to shift from cod would have happened at all.

Sailing for New Masters

After 1976, for Hull distant-water trawlermen wanting to continue in fishing, there were opportunities to work for new masters in the British inshore and middle water fisheries. During a nine-month period of unemployment, Gill's husband John took work aboard two skipper-owned small boats fishing from Scarborough. As Gill recalls, the job required cultural adjustment. Trips were short, between seven and fourteen days, and as a family-based coastal operation, fishing took place around family commitments. This did not suit John, used to the isolation and independence of the Arctic. 'They'd go out in the North Sea,' Gill explains, 'but half the time they was coming back – they was radio-ing him … there was always a problem at home and they'd come back. And he went [*said*] – I can't be doing with that, so he left.'[53]

For Kevin, a transition from distant-water trawling to seine netting proved a happier experience. During the difficulties of the late 1970s, Kevin recalls the moment that he decided to transfer to the Boston Company's Hull fleet of eight seine netters, the only small boats working from the

52. Interview with Alan J.
53. Interview with Gill.

port. In the fish dock tradition, it was a friend who introduced the idea. 'I was walking off the dock one day,' Kevin narrates, 'and [*my mate had*] landed 220 kit and he had like 400 quid to pick up, cos it was prime fish. And I was walking off the dock with about 27 quid for landing 3,000 kit. And I went [*said*] – that's it, I'm off [laughing].' Kevin followed the seiner fleet when it was sold to Consolidated Fisheries in Grimsby and continued in this line of fishing for the next 30 years. Kevin preferred the small boat system, where he worked under a share arrangement and felt more in control of his working life. 'When you went seine netting,' he concludes, 'you were your own gaffer.'[54]

The move to new companies could have unpredictable outcomes. Following the creation of the EEC common pond, a practice emerged of foreign owned vessels re-registering in Britain, enabling them to fish on UK quotas whilst still landing in their native port.[55] The UK's 'liberal registration arrangements' made the country an easy target for such 'quota hoppers'.[56] Throughout the 1980s Britain tried to halt the practice. The Spanish were not the only culprits, but as Spain awaited entry into the EEC, a 'flag of convenience' regime increasingly brought Spanish vessels to UK shores. With them came employment opportunities for redundant British trawlermen. Hull deckhand John K. had fished for one year from Lowestoft and had subsequently been made redundant from a shore job at a local tannery, when he took up the chance to fish for a Spanish company. He explains, 'it was people talking again in the [*Hull*] pubs. And then there was a man coming round after people and I signed and then they picked me up and we went to Ireland to join the ship.'[57] Spanish ships needed a percentage of British crew to allow them to work in UK waters.[58] Hull offered a promising recruiting ground. John fished in the English Channel and Irish Sea, sometimes landing at UK ports, sometimes at La Coruna in Spain. However, it became apparent that John's wife at home was not receiving his wages and in nearly a year of working with the company, John was never fully paid. It was an experience shared by others working for

54. Interview with Kevin.
55. *WF*, April 1983.
56. *WF*, September 1987, p.27.
57. Interview with John K.
58. Based on information conveyed in a research interview, Loader suggests that Spanish trawlers required two British crewmen to enable them to fish in UK waters. Source: J. Loader, 'The Decline of Hull's Fishing Industry in the Last Quarter of the 20th Century', Unpublished MA Thesis, University of Hull, 2010.

such firms. On taking a chance away from Hull, John had been forced to sacrifice the security of recognised and dependable firms. On this occasion there was a price to pay. He reflects, 'I was one of the unlucky ones. I was in a ship where you got robbed. The others, some of them got paid,' but for John, he laments that he 'didn't get anything actually, hardly, no'.[59]

The Gift of Oil

> God has given Britain her best opportunity for one hundred years in the shape of North Sea oil.
> Prime Minister James Callaghan, 1977

In home waters, alongside efforts to secure new long-term fishing opportunities, there arose a promising new prospect, not just for the ailing fishing industry, but for Britain's waning maritime economy. From the mid-1960s, Britain experienced a 'third industrial revolution' with the discovery of gas and oil in the North Sea. By 1967, gas was being produced off the east coast of England, followed in 1975 by oil from the eastern Scottish coast.[60] The benefits of these discoveries to Britain's sagging maritime industries were potentially substantial. Jamieson, examining Britain's shipbuilding and shipping companies and their involvement in the construction and operation of oil rigs and platforms, found their contribution to be small. However, one area where the nation gained greater success was in the provision of services to the offshore infrastructure.[61] It is here that distant-water trawlers and trawlermen, made redundant by the contraction of one seagoing industry, found roles in an expanding maritime boom.

The offshore gas and oil industry developed rapidly in the 1970s, encouraged by the oil crises, which elsewhere were compounding the problems of the British distant-water fishery. To support the expansion, a fleet of vessels was required for platform supply, safety and standby work, anchor handling, surveying, diving support and pipe laying.[62] Yarmouth and Lowestoft in the south and Aberdeen and Peterhead in the north became important offshore support centres.[63] These ports, like Hull, had

59. Interview with John K.
60. A. Jamieson, *Ebb Tide in the British Maritime Industries: Change and Adaptation 1918–1990* (Exeter: University of Exeter Press, 2003), pp.142–146.
61. Jamieson, *Ebb Tide*, pp.146–154.
62. Jamieson, *Ebb Tide*, p.146.
63. Jamieson, *Ebb Tide*, p.116.

suffered from the changing dynamics of fishing. For a nascent gas and oil industry in need of vessels, such ports had ships to spare. A history of Aberdeen fishing firm George Craig & Sons records that in the mid-1970s, companies were 'desperate to grab just about anything that floated' to support their offshore establishments.[64] In the long term, vessels specifically designed as support craft were brought into commission. In the meantime, the rugged distant-water trawlers, built to withstand treacherous seas, fitted nicely into the gap.

In common with Lowestoft and Aberdeen companies, Hull's trawling firms took the opportunity to offer laid-up vessels for charter. At the time of its sale in 1983, the Boston Company, working across several ports including Hull, was operating 27 converted trawlers in the North Sea gas and oil fields as Boston Offshore Safety Ltd.[65] The Marr Group was particularly energetic in their engagement with the new sector. A number of older ships were transferred to standby roles, yet it was in the field of exploration that the company excelled. Following the charter of their trawler *Criscilla* in 1976 for an oil exploration survey north of the Shetlands, the company went on to acquire, convert and manage a growing fleet of survey vessels under the subsidiary of J. Marr Shipping Ltd.[66] Following their success in oil exploration, the company went on to undertake varied seismic and scientific survey work all over the world. With significant investment, the fishing company Marr, just like its Aberdeen counterpart the Craig Group, turned a stop-gap foray into oil into a major new facet of their business.[67] Andrew Marr reflects that the converted freezer trawler survey ships were soon 'enjoying better returns than in the heyday of fishing'. Furthermore, it provided 'a great leap forward for us as other companies were going out of business'.[68]

Just as in their fishing ventures, the Hull companies that diversified into oil rig support took along their Hull crews. In the early days, trawlers chartered for oil rig service were able to remain in the membership of the HFVOA. It was agreed in 1975 that if a suitable 'call' was paid by the vessel

64. J. Cresswell, *Time and Tide: The Story of the Craig Group* (Aberdeen: The Craig Group, 2003), p.43.

65. B. Parkes, *Trawlings of a Lifetime* (Penrith: Cumberland and Westmoreland Herald, 1991), p.238; Thompson, *Stern Trawling Fleet*, p.20.

66. Thompson, *The Marr Story*, pp.51–61.

67. Interview with Alan J.; Cresswell, *Time and Tide.*

68. Interview with Andrew Marr.

owners, fishermen would retain their entitlement to Association benefits such as holiday pay, life insurance and pensions, just as if they were fishing.[69] The fortuitous discovery of the North Sea gas and oil fields, just as British distant-water trawling faced decline, is widely observed. Former skipper Ken reflects that 'you could sit and think that if [*the oil*] industry wasn't there, it would've been a complete disaster. We would've had something like four, five thousand men around with nothing to do … so it really was an answer to everybody's problems.'[70] Former trawlermen became numerous amongst those working in offshore support roles. Victor observes that during his time engaged within the sector, crews were often composed of 'refugees from the fishing industry'.[71] Not all Hull trawlermen remained with Hull companies. Some took work with other British or foreign firms working in the North Sea. Others went to work in the oil industry worldwide. Oil did not provide only for fishermen. Following redundancy in 1980, shore-based radio technician Ben overcame seasickness to return as a ship's radio operator for the Marr survey fleet, before eventually settling to a career aboard the North Sea rigs themselves.[72]

For the oil industry, trawlermen were a welcome addition. It was not just the ships, but the crews too that could endure the difficult environment of the North Sea. In his post-fishing role with the WFA, Ken would converse with oil industry representatives. 'I don't know how many times that I heard oil company owners, managers, saying how pleased they were with fishermen,' he reflects, 'because they had no problem, they would be out there in any weather. They were hard working.'[73] Trawlermen were engaged in several areas of offshore support; standby vessels, anchor handling, supply ships, surveying and pipe laying. There are striking differences in the remembered experience within each sector. The older sidewinder trawlers were well-suited for standby work. Yet whilst standby contracts provided jobs, they did little to deliver job satisfaction. Hard working fishermen struggled to adjust to a new pace of life. In the 1960s, when considering the appropriate length of rest breaks aboard a trawler, the

69. HCA (HC) Ref: C/DBTH/15/39. HFVOA. Item MC/29/75(a)(i) in Minutes of a Meeting of the Management Committee on 26 March 1975 and Agenda Item 2(a) for a Meeting of the Management Committee on 23 April 1975.
70. Interview with Ken.
71. Interview with Victor and Pauline.
72. Interview with Ben.
73. Interview with Ken.

Holland-Martin Safety Report had rejected union calls for 12 continuous hours. It concluded:

> fishermen are hunters who would resent long periods of confined idleness at sea, and a break as long as twelve hours in the confined conditions on a trawler would not only be longer than is needed for recovery from fatigue, but might well also in current circumstances create a problem of boredom.[74]

It is precisely the problem of boredom that is recalled by the trawlermen that went to work on standby ships. Even those that did not go themselves recall such reactions from their former colleagues, firmly embedding the matter within the realms of collective memory. The requirement for standby vessels resulted from the collapse of the Sea Gem oil rig in 1965 with the loss of 13 lives.[75] Thereafter, the role of the standby vessel was to be on hand to rescue or evacuate from an oil platform in the event of difficulties. Deckhand Bill went to work on standby for a former fishing company based in Lowestoft. He describes his experience:

Bill:	A lot of us went there … Some of the lads who I knew was out there, so I sez, I'll try it.
Jo (Researcher):	And how did you find that?
Bill:	Boring! Absolutely boring. Believe me, I've never been so bored in my life. Just sat there waiting … not for summat to happen. I mean, [*we*] used to sling a dummy off the ship, used to go pick it up, bring it back. That was just going up and down, up and down. That won't for us. We was used to work. We all the time was working – catching fish, putting it away. That was earning money.[76]

Bill left after a year due to an injury, but he doesn't feel that he could have endured it for much longer. Trawler skipper Bob too, who also worked on standby from Lowestoft, remembers it as boring, contrasting

74. Holland-Martin, *Trawler Safety*, p.63.
75. Dukes Wood Oil Museum, www.dukeswoodoilmuseum.co.uk/sea-gem.htm (accessed 30 July 2014).
76. Interview with Bill.

it to the non-stop, day-and-night activity of fishing. However, he felt that the standby companies looked after their seafarers and the job offered stability. Bob stayed on standby vessels until 2004. He describes a process of adjustment and acceptance, demonstrating a new type of endurance in the face of less stimulating work. 'Fishermen were too hard working to be able to put up with it,' he reflects, 'but the more you put up with it and you knew there was no work anywhere else, you just did it and did it and did it.'[77]

Laurie and Victor went into the offshore support sector, this time undertaking supply and anchor handling roles. Laurie describes himself as 'a North Sea taxi driver'. He missed the autonomy of commanding a fishing vessel. 'When you was in charge of a trawler,' he says, 'you was in charge of the trawler'. Working on a charter, you had to go where you were told and he concludes, 'I didn't like it at all really.'[78] Victor, however, enjoyed the work and continued for 20 years until retirement.[79] The tug-like anchor handlers, built to tow and manoeuvre the anchors of the oil platforms, were ideal vessels for ex-trawlermen well-used to manipulating gear, wires and blocks.

Another area where gear-handling skills were in demand was aboard survey and pipe-laying vessels. As a trawler skipper, Michael had been working for Marr on several post-Cod War projects overseas when he was called back to Aberdeen to retrain as a skipper for dynamic positioning vessels: ships with technology to maintain a precise course in deep waters. With his new certificate he was able to command pipe-laying and survey ships, which were often converted stern trawlers. His first big job was a North Sea pipeline from Norway to Belgium, and this was followed by work on exploratory seismic surveys. Both tasks demanded the ability to manoeuvre a ship whilst towing heavy and expensive gear; actions that were second nature to the trawler skipper. Michael recalls a new type of working partnership, with onboard scientific teams directing the technical procedure whilst drawing upon the skills of the skipper to ensure safe implementation at sea. As Michael reflects, 'The people you're working with on the ship, they're all university lads or ladies, who're not used to wrecks on the seabed.'[80]

The discovery of North Sea gas and oil was indeed a gift to the struggling British distant-water fishing fleets. In time, and specifically in

77. Interview with Bob.
78. Interview with Laurie, July 2013.
79. Interview with Victor and Pauline.
80. Interview with Michael.

the wake of the 1988 Piper Alpha oil rig disaster, more specialist offshore support vessels were introduced.[81] However, in the decade after 1976, it was the robust ships and crews, the 'refugees' of Britain's distant-water industry, who stepped into such roles. The new jobs offered relative stability and a reliable living. But they also required adjustment. In the standby sector in particular, fishermen who once pursued the big catch had to accept a role in which they hoped their intervention would not be necessary. The result was boredom. However, other jobs were more engaging and for all those working in the North Sea there was the chance of more generous breaks at home. For those engaged in exploration, the situation was different. For Michael, working for Marr in the North Sea was just the beginning of a series of survey contracts that eventually took him to places as distant as Vietnam, America and India.[82] After the Cod Wars, rather than move into home waters, Michael became part of a cohort of Hull trawlermen and trawling firms who, in seeking out new opportunities, increasingly found themselves scattered all over the world.

Far-flung

A Globalising Industry

The South-West England mackerel fishery had brought Hull trawlers into home waters, focusing for the first time upon overseas distribution and disrupting the long-established dependency upon the home port and home markets. After 1976, Hull companies increasingly operated within international parameters, and men of the former Arctic trawl found themselves spread worldwide in what was fast evolving into a globalised fishing industry. The earlier limited focus and domestic markets of the traditional Hull fishery had not been unique. Wilcox observes that early constraints in technology and available markets had tied many Northern European fishermen to the North Atlantic and to landing at their home port.[83] The years following the Second World War saw the spatial expansion of European activity. From the 1960s, the need for Eastern bloc states to secure cheap protein caused them to become prolific fish catchers. Drawing upon freezer technology, they

81. Cresswell, *Time and Tide*, pp.44–45.
82. Interview with Michael.
83. M. Wilcox, 'Beyond the North Atlantic', in Starkey and Heidbrink, *History of North Atlantic Fisheries*, p.307.

extended their reach beyond the North Atlantic into the South Atlantic, Africa and the Pacific. Mediterranean nations also increased their catching effort in the waters of Africa and the Indian Ocean.[84] However, in Northern Europe, Britain and Germany continued as before. Until the 'pull' of new opportunity was compounded by the political 'push' of fishing limits, there did not seem to be a need for change.[85]

The threat posed by technologically advanced and roaming fishing fleets undoubtedly spurred the rapid expansion of 200-mile EEZs. By 1979, the number of EEZs had increased to over ninety worldwide.[86] The EEZ gave control to the coastline nation to manage its own fishery efficiently, so to avoid overfishing and collapse. For countries wishing to fish in another nation's waters an agreement had to be reached, which generally involved a financial or other reciprocal benefit for the coastline state. For European members of the Common Market, such negotiations were the responsibility of the EEC. Although EEC negotiations initially focused upon maintaining access to the traditional grounds of member states, from the late 1970s, agreements were expanding geographically and sometimes controversially, with the aim of supporting European markets and keeping excess European fishing capacity at sea.[87] EEZs provided an opportunity for nations with small-scale fisheries to expand their industry. In joint ventures, developed countries could offer developing fisheries the chance to draw upon the skills and experience of established fishing nations. For developing nations, overseas partnerships could bring aid or payment. Partnerships could also secure the resources needed for growing processing sectors or could secure access to overseas markets in an increasingly globalised fish export economy.

As a leading fishing nation, Britain's distant-water industry potentially had much to offer and to gain by entering into overseas ventures. Some have criticised the limited level of overseas engagement that was undertaken by UK distant-water firms. In a 1986 interview with Euro MP James Provan, a *World Fishing* journalist suggested that with the exception of Spain, other European distant-water fishing interests, including the UK, had shown little entrepreneurial spirit.[88] Provan agreed that the UK fleet had indeed

84. Wilcox, 'Beyond the North Atlantic', pp.307–309.
85. Wilcox, 'Beyond the North Atlantic', pp.309–310.
86. *WF*, October 1979, p.5.
87. A. Le Sann, *A Livelihood from the Sea: Globalisation and Sustainable Fisheries Policies* (London: Intermediate Technologies Publications, 1998).
88. *WF*, February 1986, p.5.

missed opportunities. There had been, he says, 'plenty of fishing available off Canada, off the Americas, etc., if they had been prepared to go and do it, and they weren't; they would probably have done well'.[89] Peter, who emigrated to Canada in the early 1970s for a fisheries management position, shared his own disappointment that Hull trawling firms did not follow, but instead preferred to 'chase the same old thing'.[90] For Northern Europeans steeped in a home port tradition and only beginning to come to terms with territorialisation and a global arena, the move towards worldwide operation required a massive shift in outlook. There were also practical hurdles to overcome. Overseas ventures were far from stable. They required substantial investment and, in his article, Provan does concede that the British government had offered little financial support to secure the right conditions for entrepreneurship.[91] Furthermore, UK involvement with joint ventures could result in unwanted concessions, such as a reciprocal influx of overseas fish to depress home markets.

In Hull, another factor had anchored the fleet to home. For decades, trawler owners had built up dockside services, facilities and subsidiary businesses located in the home port, to serve and profit from the North Atlantic trawl. The divergence of the surviving fleet away from the city would put these facilities and the resources invested in them under pressure. Earlier, Peter reported his impression that trawler owners could not conceive of their ships not returning to Hull. From the mid-1970s, the mackerel fishery had seen trawlers beginning to engage less regularly with the port. Operating away from the city had an impact in the struggle to keep the home port and its investments financially viable and perhaps this contributed to the reticence to engage more fully overseas. A shift to more nomadic operations might be costly in business terms and would require a rapid and substantial adjustment in outlook and practical organisation. Furthermore, it would require the psychological leap of relinquishing the fish dock, and the structures and dependencies that had accrued over a century of trawling.

Despite constraining factors, Hull trawling firms did engage overseas in the difficult years after 1976. Their involvement never matched the zeal of nations such as Spain or the states of the Eastern bloc. Yet at times it was not without vigour. Through EEC joint ventures, private partnerships and

89. *WF*, February 1986, p.5.
90. Interview with Peter.
91. *WF*, February 1986, p.5.

the sale of vessels, Hull companies sought new horizons and sent their ships to fish off unfamiliar coasts. With them went Hull crews, keen to work and willing to try new fisheries in warmer climes. For those unable to fish with Hull firms, just as in home waters, there were opportunities to work directly for overseas companies. It was not always plain sailing and in the new global environment it was wise to proceed with caution. Oil also offered worldwide possibilities for former trawlermen, just as it had in the North Sea. In the years after the Cod Wars, the tight-knit networks of Hull's fish dock went global, as the men of the Northern trawl plied their trade and skills in increasingly distant seas.

Leaving Home

For those willing to leave the UK, there were multiple overseas openings. Some proved to be short lived. As a trawler skipper, Bob had a brief sojourn in Iceland, waiting with another Hull skipper at Partisfjord to take command of new trawlers for an Icelandic company. As Iceland built up its fishing capacity, Bob explains that firms wanted to benefit from the knowledge of Hull skippers. Hull trawlermen had worked Arctic grounds over decades and Bob contends that such men knew 'more about Icelandic waters than the Icelanders did'.[92] However, the new vessels were not to materialise. Bob spent his time working in the company's fish salting factory, with the occasional trip on the firm's existing ships. After 11 months in Iceland, when the ships failed to appear, Bob gave up and returned home to standby work in the North Sea.[93]

For others, working overseas became a more permanent career. Michael's narrative illustrates this. Following his first trip transporting two Marr trawlers to new African owners, Michael's career took a predominantly overseas pathway. Often but not always for Marr, he worked in the waters of Africa, America, Asia and occasionally Europe, staying away from home for months at a time. For those working in association with Hull companies, the length of time spent abroad could depend upon the success of the venture. Hull firms became involved in projects that had their origins in both private negotiations and government agreements. From 1979, when the West African coastal state of Mauritania revised its fishing policy to improve control of foreign fishing effort, Hull firm Thomas Hamling &

92. Interview with Bob.
93. Interview with Bob.

Company was amongst British companies seeking joint ventures under the auspices of the EEC.[94] The venture was short lived, however, as the company went into receivership in 1983. In 1977, with Australia keen to establish its own deep-sea fishing industry in advance of declaring its own exclusive 200-mile zone, the UK firm British United Trawlers joined forces with Australian company Southern Ocean to establish an integrated catching, processing and marketing operation.[95] Three Hull freezer trawlers, *Othello*, *Cassio* and *Orsino* made their way to Albany in Western Australia, along with crews consisting mainly of Hull men.[96] It was planned that the nascent Australian industry would benefit from the experience and skills of British management and crew, whilst Australian fishermen were trained to take over in the future.[97] However by May 1979, the partnership had gone into receivership. The Hull crews found themselves at the centre of a dispute over who should bring them home, as negotiations commenced to sell on the ships to Nigeria.[98]

In New Zealand, the reluctant sale of the *Arctic Buccaneer* and *Arctic Galliard* had a happier outcome. Like Australia, New Zealand was keen to establish a deep-sea fishery in their newly established 200-mile EEZ. Bombarded with joint venture applications from the sizeable fleets of Russia and Japan, yet faced with the political bartering of fishing rights against dairy markets by those countries, New Zealand had actively encouraged the involvement of the British distant-water industry. But in 1977, the terms being offered were unattractive to British firms.[99] It was not until 1982, with the enforced sale of his cutting-edge freezer ships that Tom Boyd was to engage with New Zealand, as Boyd Line continued to manage the vessels on behalf of the new owners, Fletcher Challenge. The ships sailed with a mixed crew of Hull trawlermen and New Zealand trainees, as the nation sought to build up its own domestic deep-sea sector.[100]

Leaving home for the South Pacific was a big decision. Tom recalls the moment that he broke the news that the ships had been sold:

94. D. Gibbs, 'Politics of Economic Development: Case of Mauritania Fishing Industry', *African Studies Review*, 27:4 (1984). HCA (HC) Ref: C/DBTH/6/4/11. Various Papers Relating to Hamling's Interest in West Africa, including Mauritania, 1979–80.
95. *FN*, 27 May 1977, p.3.
96. *FN*, 18 November 1977, p.4.
97. *FN*, 27 May 1977, p.3.
98. *FN*, 22 June 1979, p.3 and 5 October 1979, p.1.
99. *WF*, September 1977, pp.3–4.
100. *FN*, 6 August 1982, p.3.

> At that time, we were briefing [*the crews*] at the end of every trip. We got the crews in and said we'd got good news and bad news for them and I was going to give the bad news first and that was that their ships had been sold … There was a horrible silence. And then we said, but the good news is that those of you who would like a job, there is a job for you in New Zealand, on these ships. And I can't tell you what you'll be paid … but we will be managing them and we will negotiate for you, whatever we can and it will be a fair rate.[101]

Despite the life-changing impact of operating so far from home, many crew members remained with the now renamed *Otago Buccaneer* and *Otago Galliard*. With reducing opportunities at home, Hull trawlermen wishing to remain in distant-water fishing had to broaden their horizons. And as one Boyd manager commented, New Zealand offered the chance to 'see something building up rather than running down'.[102] The New Zealand venture was a success, as, freed from the restraints of small quotas, the performance of the vessels and experienced Hull crews turned out to be exemplary.[103] Fishing proved lucrative and Boyd Line continued to manage the operations for the following decade, effectively kick-starting the New Zealand deep-water industry.[104]

Long-term opportunities for former Hull trawler skippers were also available in the overseas oil industry. Just as in the North Sea, the skills of the skipper were welcome in the global offshore support and supply sector. Crews however, tended to be local, or with the maritime shift towards globalism, international. In his autobiography, Hull skipper Danny Platten described how his move to Africa with offshore support company OIL was spurred by his knowledge that other trawler skippers had done the same. He worked in the African oil industry for 14 years from 1984, enduring a tumultuous political situation in return for a steady income and a job that he enjoyed.[105] Gill's husband John also took a job with supply companies in the Gulf, remaining in the business for 19 years.[106] Ron too found work

101. Interview with Tom Boyd.
102. *FN*, 5 February 1982, p.3.
103. *FN*, 2 August 1982, p.3.
104. *FN*, 2 August 1982, p.3.
105. D. Platten, *From the Arctic to Angola: The Adventures of a Hull Skipper* (Hull: Riverhead, 2010). pp.134–139.
106. Interview with Gill.

with American–Arabian oil company Aramco in Saudi Arabia, teaching seamanship and chart skills to Saudi nationals aboard the company's own training ships.[107]

Working abroad also bought opportunities to emigrate. As seafarers, many Hull trawlermen were willing to try their luck in distant seas. As Ben reflects, at that time Hull had 'always been that sort of a town, guys [*had*] always worked away'.[108] But emigration meant disrupting families and pulling up roots. It did not appeal to all. Dave refused an offer to emigrate with a Hull fishing venture in Australia because the project could not offer a guarantee of success and stability.[109] Christine's husband Paul was offered the chance to emigrate to Canada or New Zealand, but close family ties made Christine reluctant to move and they chose instead to remain at home.[110] For seafarers like Dave and Paul, working on distant horizons meant voyaging further afield and spending more time away. But it also meant coming back. In this new lifeworld, ties to Hull may have been weakened, but family bonds often ensured that they were not broken altogether.

Scattered: Encounters Beyond the North Atlantic

For those working on overseas projects, there were new experiences and encounters. Until the mid-1970s, the operation of British distant-water fishery had confined many to a cold climate and established parameters. Michael's narrative reveals how different life could be in a global post-Cod War era. Delivering two Marr trawlers sold to Ghana, Michael sailed into unfamiliar seas. Accustomed to Arctic waters, Michael reflects on an unusual voyage. 'I'd never been further south than Grimsby in me life,' he explains, 'and I was 38, so when I saw the sunshine and all that … well I couldn't believe it. I thought, what have I been doing [laughing].' He remembers the spectacles of flying fish, white sandy beaches, palm trees and turtles. 'It was paradise,' Michael recalls. 'I thought, bloody 'ell, this is the life. We've arrived.'[111]

Michael stayed in Ghana for a year, first helping with the handover to local crews and then working directly for the Ghanaian company. Here

107. Interview with Ron.
108. Interview with Ben.
109. Interview with Dave.
110. Interview with Christine.
111. Interview with Michael.

again, there were unexpected protocols. Michael was surprised when the crew were required to fish for their own sustenance on the journey to the fishing grounds, something that would never have been requested on the run out of Hull. Others similarly encountered new cultural experiences when working overseas. Ron describes the privileged lifestyle he commanded as a British worker in an American–Saudi company. As a senior staff member Ron was provided with attractive sea-front accommodation and he says, 'It was like being on holiday all the time.'[112] Working in the Gulf, Gill's husband John skippered Indian or Filipino crews. In his spare time, he would ask the ship's cook to teach him how to make authentic curry dishes.[113] Ben, after working for Marr in Angola, Cameroon, Singapore and Indonesia, found himself employed by an American seismic company and experienced living in a houseboat moving around the Delta and tying up in the Mangroves.[114]

Sometimes it was not just about adapting to new places, but to new roles. Michael tells stories from his time as the skipper of a small cruise vessel sailing around the Falkland Islands. On one occasion, Michael summoned back his passengers from a visit to an Island farmhouse, concerned by a swelling sea in the rocky cove where he was anchored. The tourist rep called the farmhouse, and Michael recalls:

> I'll never forget, I was really worried, all these rocks and that … [*the farmer's wife*] said, I've poured the tea out, all cups and saucers. So [*the rep*] said, what shall I tell her? I said, you can tell her what the bloody 'ell you like, they're not going and that was that.[115]

More than an anecdote, the story illustrates how different life could be in the days after the Cod Wars. Negotiating the sensibilities of paying guests and tourist providers could not be further from the life of the independent trawler skipper with an absolute command of ship and crew. In going to work overseas, it was not only physical, but sometimes metaphorical distance that had to be navigated.

Work overseas also brought ex-trawlermen into circumstances that were politically unstable, dangerous and corrupt. Those working for Hull

112. Interview with Ron.
113. Interview with Gill.
114. Interview with Ben.
115. Interview with Michael.

companies had the relative stability and protection of known and reliable firms. Although British United Trawlers had wrangled over assisting their crew members stranded in Australia, in Ghana, Michael was confident, even when working directly for the Ghanaians, that the Marr company would have looked after him should a problem have arisen. Indeed Marr, which instigated Michael's involvement in Ghana, sorted out his payment when it was not forthcoming from Africa. Like John K. and his experience with the Spanish fishing company, Michael was aware that 'a lot of Hull skippers had been to Africa, different countries and places and never got paid'.[116]

For those working directly for overseas employers, the situation could be precarious. Projects to expand fisheries in developing nations could be disorganised. Danny Platten tells how he ended his management contract with Guyana Fisheries as the company suffered constant technical and staffing difficulties.[117] Worse problems could be posed by the political environment, by cultural protocols and by unreliable employers. In July 1979, the Hull Trawler Officers' Guild was celebrating new opportunities for its members as three Hull skippers flew out to Nigeria to contracts with the American company Sea State Seafoods. The Guild explained that 'all those going abroad want to keep in the fishing industry and are prepared to go anywhere to follow occupations'.[118] Yet within one month, all three were rushed home as the American company director was inexplicably threatened with house arrest and the ships with impoundment.[119] Michael, in his prolific and varied overseas career encountered corrupt management, drugs amongst the crew and guns stored aboard.[120] Even when working for stable companies, the environment itself could be dangerous, as Danny illustrates in his account of evacuating onshore oil personnel out to nearby rigs during fighting in Angola in the 1990s.[121] In the past, fishing in the Arctic had brought its dangers, but it had operated within recognised boundaries. In the new global arena, working overseas could mean a journey into a hazardous unknown.

Whether working for reliable companies or grasping at insecure contracts, the decade after 1976 saw Hull trawlermen scattered across the

116. Interview with Michael.
117. Platten, *Arctic to Angola*, p.133.
118. *FN*, 13 July 1979.
119. *FN*, 10 August 1979.
120. Interview with Michael.
121. Platten, *Arctic to Angola*, p.158.

globe. As the Trawler Officers' Guild had emphasised, men wanted to keep fishing and many were willing to go to great lengths to do so. For some, the Hull connection was maintained as they continued to work with Hull firms and crews on very distant shores. Yet even when working for new overseas employers, the skills that Hull trawlermen held in common often caused their paths to cross. In fact, the continuance of former fish dock networks actively contributed to Hull men securing jobs in the same locations. Ron regularly met Hull trawlermen in Saudi Arabia, as they skippered diving support ships and supply vessels. He remembers one such encounter walking along the sidewalk among the Arabian shops. 'All of a sudden I heard a voice shout from across the road "Now then Ron!"' he recalls. '[*I thought*] who the hell is that? And it was a kid [*man*] who I knew from the fishing industry. I thought to myself, good grief, I'm in the middle of Saudi Arabia and all of a sudden you meet somebody from Hull.'[122] Gill's husband John also met Hull trawlermen in the Gulf oil industry. Reminiscent of his time as a skipper on the Arctic fishing grounds, encounters often happened over the ship's radio and Gill recalls John's observation: 'he said, it's like being at Iceland or Bear Island or somewhere, you wanna see the blokes I've met out here, y' know, saying all their names.'[123]

With the need to recoup losses, it was not just trawlermen, but trawlers too that were dispersed worldwide, as they were sold on to new owners. Sometimes former Arctic ships and crews came face to face on distant shores. Michael recalls seeing *Junella*, Marr's first whole fish stern freezer trawler and famous in its day, along with two Fleetwood trawlers, moored in South Africa. 'It was just past us … tied up in Cape Town,' he reflects. 'That's a million to one chance.'[124] Alan J. had a similar encounter in St John's, Newfoundland. 'I was just looking out the hotel window,' he recalls, 'and – that's an ex-Hull trawler there – and it was. And [*you'd*] see them all over the place.'[125] After the Cod Wars the intense local focus of the Hull trawl fishery was diluted. Although bonds remained, the home port became increasingly remote, in relevance and distance, as ships and men scattered across continents. To Michael the sight of *Junella* in Cape Town said it all: 'It just shows you,' he reflects, 'how we was spread out all over the world.'[126]

122. Interview with Ron.
123. Interview with Gill.
124. Interview with Michael.
125. Interview with Alan J.
126. Interview with Michael.

Developing the Falklands: A New British Fishery

From the late 1980s, there arose a major opportunity to develop a focused, comprehensive and sustained new fishery that this time would be in direct British control. In 1976, an Economic Survey of the Falkland Islands undertaken by Lord Shackleton proposed the development of a new South Atlantic fishery as a means of realising the economic potential of the Islands. The purpose of the survey had been to assist the flagging social and economic conditions on the British outpost.[127] The resulting recommendation for a Falkland Islands 200-mile exclusive fishing zone presented an opening, just as the UK distant-water fleets were faced with the closure of the North Atlantic.

By 1977, consultation was underway with the BFF.[128] By 1982, members were offering their services for the establishment of a South Atlantic fishery, with the Federation asserting that they alone in the UK 'had the commercial fisheries expertise that would be relevant, but that whatever was done would need to be underwritten by HMG'.[129] The economic climate of late-1970s Britain, followed by the Argentinian war over sovereignty of the Islands in 1982, served to delay progress. Furthermore, the ensuing strained relationship with Argentina and her allies in the years after the UK victory meant that the multilateral conservation and management agreement desirable as part of an EEZ declaration stood little chance of fruition.[130] In the meantime, the area outside the Falklands' meagre three-mile limit was being fished relentlessly. Fleets from the Eastern bloc, Spain, Japan, Taiwan and South Korea took advantage of the absence of restrictions to fish freely and extensively upon squid, hake and blue whiting, with some openly declaring their intention to fully exploit the bonanza until the UK government decided otherwise.[131] With little cooperation from Argentina in the conservation of a shared resource, fishing continued and the risk of stock depletion threatened to

127. Falkland: Our Islands, Our History, www.falklands.gov.fk/assets/OurIslands OurHistory.pdf (accessed 11 August 2014).
128. HCA (HC) Ref: C/DBTH/20/73 BFF. Item 77/181 in Minutes of a Meeting of the Board of Directors on 6 October 1977.
129. HCA (HC) Ref: C/DBTH/20/82. BFF. Item 82/66 for Minutes of a Meeting of the Board of Directors on 29 September 1982.
130. *WF*, December 1986, p.3.
131. *WF*, October 1984, p.10.

wipe out Britain's newest fishery before it had even begun.[132] Eventually, in October 1986, Britain announced a 150-mile interim fishing limit around the Falkland Islands, with fishing licences issued from February 1987.[133] In the initial years of the fishery, revenue came from the fees of a managed system for licensing foreign vessels. However, those applying for a licence were strongly encouraged to enter into a joint venture with Stanley Fisheries Ltd, the company created by the Falkland Islands Development Corporation with the aim of giving Islanders a direct and long-term stake in their own fishery.[134]

For Hull trawling firms, the Falkland Islands presented a chance to apply their skills and experience to a developing fishery, which under the rules of Overseas Countries and Territories enjoyed a degree of autonomy from EEC Law and the CFP.[135] J. Marr & Son and Boyd Line, the two remaining Hull distant-water companies in the late 1980s, embraced the opportunity with energy and enthusiasm. In April 1987, *World Fishing* praised the support of Marr in the establishment of the Island fishery.[136] Five Marr freezer trawlers had been requisitioned as part of the Falkland's Naval Task Force during the 1982 conflict. In the ensuing years the company had continued to promote to government the benefits of a fishing zone around the Islands, laying out detailed suggestions to tackle the practicalities of such a venture.[137] With the establishment of a 150-mile zone, Marr, along with Boyd Line, threw themselves into a range of operations that could take the full advantage of the development.

In the ten years since the initial collapse of Arctic trawling, both companies had learned new ways to fish. They engaged with the Falkland squid fishery by entering into joint ventures with Japanese squid-jigging vessels.[138] Such ventures allowed British trawling firms to gain experience in new techniques. However, the revision of licensing arrangements by the Falkland Islands Government eventually removed this type of partnering opportunity.[139] Tom Boyd describes his additional disappointment as

132. *WF*, November 1986, p.3 and December 1986, p.3.
133. *WF*, December 1986, p.3 and April 1987, p.2.
134. *WF*, April 1987, pp.2–3.
135. J. Ziller, 'The European Union and the Territorial Scope of European Territories', *Victoria University of Wellington Law Review*, 38 (2006).
136. *WF*, April 1987, p.2.
137. Interview with Andrew Marr and *WF*, October 1984, pp.10–11.
138. *WF*, November 1986, p.33 and April 1987, p.3.
139. Interviews with Andrew Marr and Tom Boyd.

permission was also withheld for the launch a potential salmon ranching project on the Islands. His experimental venture to catch and process crab, partnering with leading UK crabmeat firm Smirren's, similarly met with restriction.[140] The outcome was frustrating and the costs for Hull companies fishing in the Falklands was becoming prohibitive.

Both firms also operated their own vessels in the South Atlantic. Boyd had the *Arctic Freebooter*, renamed the *Lord Shackleton*. Marr had *Junella*, which following an expensive conversion became *Hill Cove*, owned by Stanmarr, the joint venture between Marr (Falklands) Ltd and Stanley Fisheries. In 1988, *World Fishing* observed that Hill Cove was leaving 'one of the world's oldest fishing communities, Hull, to join one of the newest, Port Stanley'. Onboard was a skipper and crew from Hessle Road.[141] In many ways the *Hill Cove*, designed to fish flexibly on species found in the Falklands, was a final departure from the cod dependency of a decade earlier. Yet diversity could not protect the venture from a volatile market. Nor could it avoid legislative changes that ultimately led to the dissolution of the Stanmarr partnership just a year later, with the subsequent sale of *Junella* to Uruguay.[142]

Although direct fishing ventures in the Falklands faced difficulties, the two firms had greater success in managing the fishing limit. With their longstanding experience gained in distant waters, they were in a position to understand and control fishing operations. From 1987, J. Marr Shipping took on the major role of fisheries protection, providing vessels and crews for the Falkland Islands Patrol Service with their ships, *Falklands Desire*, *Falklands Right* and later *Falklands Protector* and *Cordella*.[143] Michael was amongst those who worked for the company in the South Atlantic. Michael had openly breached the Icelandic limits during the Cod Wars, but now he chased unlicensed Taiwanese ships towards South Africa and halted the painting of false registration numbers onto Japanese vessels.[144] Engineer Dave also joined the fishery protection crews in the Falklands, seeing it as a means to pay off his mortgage. In the years after the Falklands War, he remembers being flown from RAF Brize Norton, acquiring a fighter plane escort as they approached the Islands, and being thoroughly briefed about areas that were off limits due to

140. Interview with Tom Boyd.
141. *WF*, March 1988.
142. Interview with Andrew Marr and *WF*, July 1989, p.21.
143. Thompson, *The Marr Story*, p.61.
144. Interview with Michael.

landmines.[145] Ben too worked as a radio operator on the Marr protection vessels. He recalls a patrol system in its infancy. Obtaining information from the foreign vessels, he says, was 'like pulling teeth', but generally it was amicable.[146] Although the fishing ventures of Hull companies in the Falklands were short lived, the protection role proved more enduring. Precarity nevertheless remained. Dave recalls the time when both his trip and his seagoing career were brought to an end. It was his second trip aboard *Falklands Desire*, known within Hull as *Southella*. *Southella* had been Dave's first ship when started out in trawling, and it was to become his last. When the Falkland Islands Government no longer needed as many protection ships, one had to go. 'It just happened to be ours,' Dave reflects, 'so we sailed up to Cape Town, disembarked there, stamped my book … and that's my last trip to sea'.[147]

The trawling firms Marr and Boyd played a substantial role in the South Atlantic. In 1987, as the long-anticipated managed fishery got underway, *World Fishing* reflected: 'Marr are justly proud of their involvement in the Falkland Islands Fishing Zone and not only that they have helped to create it, but that they are taking a leading role in making it work, both from the control aspect and in fishing the zone.'[148] Their involvement was energetic, diverse and committed, with the company actively encouraging Islanders to become part of their own fishery.[149] As Marr had predicted in the early planning stage, the fishing zone brought significant wealth to the flagging economy of the Islands. In 1985/86 the income of the Falkland Islands Government was £6 million. By 1988/89 fishing had boosted this to £35 million, transforming the Islands with affluence and injecting self-sufficient investment into public services and infrastructure.[150] Over a 25-year period, the Islanders' position within the fishery shifted from one of agent, to partner and finally to fishing vessel owners in their own right.[151] There were multiple contributors to the South Atlantic achievement. Among them, were the adapting companies and experienced crews of the older Hull-based North Atlantic trawl.

145. Interview with Dave.
146. Interview with Ben.
147. Interview with Dave.
148. *WF*, April 1987, p.3.
149. *WF*, April 1987, p.3.
150. Falkland: Our Islands, Our History.
151. Falkland: Our Islands, Our History.

A Place in the World: Importing Fish into Hull

As Hull firms, ships and crews chased opportunities in distant seas, back in the home port the needs of the city's substantial merchant and processing sector were driving the development of new supply networks. As fish landings fell and the home fleet departed, the fishing port of Hull was compelled to open its gates to international markets. Fishing was swiftly emerging as a globalised industry, and the fish trades were prominent in leading this largely demand-led transformation.[152] The circumstances imposed by stagnating catches in the North Atlantic and the extension of EEZs increased the global trade in fish as Western nations turned to imports to supply domestic markets.[153] As Holm observes, 'frozen fish blocks became the basic currency unit in the fishing industry worldwide'.[154] In the UK, imports increased rapidly in the 1980s and 1990s. The high volumes of fish landed at the Humber ports prior to the crises of the 1970s had established Hull and Grimsby as major fish handling and processing centres. For over a century the Hull fish trades had been supplied predominantly by Hull vessels landing at their own fish quay. In 1970, a total of 197,014 tons of fish was landed at Hull fish dock. By 1981, in the immediate wake of limits and quotas, this had fallen by 92 per cent to a mere 15,217 tons.[155]

Maintaining supply was critical for firms making a living from the sale and preparation of fish. If fish could no longer be provided by the city's own fleet, then new suppliers had to be attracted to the port. The result was a growing autonomy for the fish trades, with the former local links between fleet, filleter and factory proving to be dispensable.[156] Fish merchants David and Chris remember the transition to national and international supply networks. It was a process that had begun prior to the final Cod War. By the mid-1960s, Hull's growth as a freezer port, coupled

152. Wilcox, 'Beyond the North Atlantic', pp.313–314.

153. P. Holm, 'An International Perspective on Britain's Fisheries in the New Millennium', in Starkey et al., *England's Sea Fisheries*, p.239.

154. P. Holm, D. J. Starkey and J. Thor (eds), *The North Atlantic Fisheries 1100–1976: National Perspectives on a Common Resource*, Studia Atlantica: 1 (Esbjerg: NAFHA, 1996), p.239.

155. SFIA, *Hull Fish Industry in 1983*, p.2.

156. In 1982 126,100 tons of fish entered Hull via all means. Of this less than a quarter was landed through the Fish Dock and of this quarter, only 7 per cent of frozen fish and 5 per cent of wet fish were landed by Hull vessels. SFIA, *Hull Fish Industry in 1983*, pt II.

with falling catch rates, had led Hull merchants to procure Scottish-caught wet fish to supplement supplies.[157] As the supply crisis deepened through the 1970s and 1980s, increased volumes of Scottish fish travelled overland into Hull, accompanied by fish from the inshore fishing ports of the Yorkshire Coast. In addition, wet fish and frozen fish was imported from Iceland, Norway, Ireland, France and East Germany.[158] Imports were packaged in containers and arrived via the Humber's commercial docks, travelling into Hull's fishing district by road. According to Chris, fish was transported via a freight ferry that travelled between Reykjavik, the Faroe Islands and the Humber. On arrival in Hull, the fish was graded, sorted and processed. 'It was still haddock, cod and whatever,' he explains, 'but the vessels didn't come to the quayside. They stayed in Iceland and it was the containers themselves that came to Hull to be processed.'[159]

Merchant David welcomed the arrival of much-needed Icelandic supplies. The new system was not without logistical problems. Container vessels landed once, perhaps twice a week, sometimes making it difficult to supply the weekend frying trade.[160] As the supply of wet fish fluctuated, there was a growing dependence upon frozen imports. David was an early convert to a frozen supply, handling fish from Hull's freezer fleet from the early 1970s. Although some of the smaller merchants lacked the facilities to deal with frozen blocks, David made contact with several Norwegian companies, eventually choosing to buy directly from a single company. Over time Norwegian frozen fish became David's dominant supply. 'That's our particular trade,' he reflects, 'and that's how we've survived.'[161]

Alongside containerised imports, fish was landed directly at Hull's fish dock by foreign 'fresher' and freezer trawlers. Many came from Iceland and there was some initial resentment amidst the struggling Hull fishing community. For deckhand Thomas, the memory of Icelandic trawlers landing in Hull in the wake of the British ban from Iceland is still emotive. 'Icelandic trawlers landing direct here! Because we couldn't fish there,' he recalls in frustration. 'It was terrible.' The landings, he says, created 'massive bad feelings, because they took our living away'.[162] Two Icelanders known to

157. SFIA, *Hull Fish Industry in 1983*, p.33.
158. SFIA, *Hull Fish Industry in 1983*, pp.34–46.
159. Interview with Chris.
160. Interview with David.
161. Interview with David.
162. Interview with Thomas.

Chris at this time had been refused service in local pubs. 'It wasn't fierce,' he explains, 'but you can understand it. If you went into a pub on Hessle Road and you were Icelandic, you wouldn't be particularly welcome, put it like that.'[163] Antagonism did not last long. Despite sensitivities, Icelandic and other foreign-caught fish were vital to Hull's fish trades and to the survival of the fish dock itself. Just as in the catching sector, international operation brought vulnerability to global negotiations and politics. In 1984, Alan J. set up his own Hull-based fish-trading company, importing Russian fish into the UK. In the early 1990s, he says a ban on the import of Russian fish into the EEC wiped out 90 per cent of his business.[164] However, the potential and the willingness to source from the fisheries of the world gave the Hull merchant and processing sector the means to retain and adapt their trade in the difficult years after 1976. The 1983 analysis of the Hull fishing industry produced by the SFIA praised the adaptability of the fish trades, concluding that 'it has been the business acumen of many firms – both large and small alike – that has secured for Hull an important share of the UK's distribution and processing activities'.[165]

In the years that followed, Hull was to establish its place in the world as a European centre for the trade of fish. However, after the Cod Wars the combination of freezer technology and a globalised trade brought a different experience to the former home fleet landings. Alan J., manager with both the catching and trading sector, recalls the switch to handling a raw material that was rarely glimpsed. 'You bought it,' he explains, 'you put it into a ship and sent the ship to a Danish port or whatever, you sent a couple of guys over there to supervise the discharge and make sure everybody was happy with the quality and that was it. You never saw it.'[166] The traditional shoreside functions, once essential to the handling of a perishable product, were consigned to the past. Fish had become a commodity just like any other. 'A block of frozen fish doesn't need any processor,' observes David. 'It just needs someone to store it and sell it. You know, it's a bit like selling boxes of cornflakes.'[167]

163. Interview with Chris.
164. Interview with Alan J.
165. SFIA, *Hull Fish Industry in 1983*, pt II.
166. Interview with Alan J.
167. Interview with David.

New Lives

In the decade after the final Cod War, life in all quarters of Hull's fishery altered forever. The dependable rhythms of trawling had dissolved. Thousands had experienced unemployment or faced social and economic upheaval. Emerging from the wreckage was a much-reduced industry working in a much-expanded arena. In this new era, fishing had become political, territorial and above all global. For those who had struggled through the transition, times had been turbulent. Big changes had imposed new lives and narratives after 1976 are a far departure from the established blueprints of the old North Atlantic trawl.

Unsurprisingly, it is within the seagoing sector that change is most discernible. For some older men riding out redundancy in pubs or on street benches, 1976 brought their seagoing and sometimes their working lives to a close. Others had decided to 'swallowed the anchor' and find permanent work ashore. Sometimes, like Ken in the WFA, the new shore job related to fishing. Sometimes, like Thomas in the army, there was a clear break with the past. For some, staying at sea was imperative. Jim, working on the Humber pilot cutters, was clear that being at sea was all he wanted to do. For those like John K., Dave, and Christine's husband Paul, new lives could fluctuate between sea and shore. John's experience highlights the instability that could follow. John's hard work ethos kept him employed, but a combination of family circumstances, redundancy and unreliable firms saw him change jobs several times before eventually retiring.[168] Despite its casual basis, with hard work and good fortune, trawling had offered a regular living and, for the ambitious, the chance to make it to skipper. In working lives after 1976, there were no such promises.

John K's narrative highlights another major change for those leaving trawling. In the heyday of the 1950s, Tunstall showed that despite the poor hourly rates, trawling offered the chance of lucrative annual incomes, far exceeding those of other working-class jobs ashore.[169] But on land or at sea, employment opportunities after the 1976 collapse rarely matched the earning potential of the big catch. For experienced skipper Ken, not even working in a specialist professional job as a cartographer could he equal his income from trawling. In the same vein, John's foray into taxi driving as he moved towards retirement proved to be a financial miscalculation. It was, he

168. Interview with John K.

169. Tunstall, *The Fishermen*, p.69 and p.207.

says, 'the worst thing I could've ever done … I was hoping to stop, cut me hours down. But I was having to work more hours than I was even when I was at sea, y' know, to earn a living.'[170]

It was not only finances that sent men back to sea. After the collapse of trawling firm Thomas Hamling & Company, Christine's husband Paul found a job that he enjoyed at a local tannery. But Christine recalls that he got 'itchy feet'.[171] She reflects, 'I know he missed it … he said he'd never go back to sea, but I knew that he would. I knew he would. It was always there.'[172] For the next 23 years, Paul secured one of the dwindling number of fishing jobs with Marr, and Christine readapted to her role as a fisherman's wife. Dave had also returned to sea on several occasions before finally coming ashore in the 1990s. With his engineering qualifications, Dave adjusted easily to shore-based employment. However, he describes a fatigue with the regularity of shore life, which had lured him back to sea in the past:

> you come ashore for a while and you want to go back to sea. It's a pull somehow. I don't know whether it's the routine ashore that you get sick of. Having to do this, having to do that. Whereas at sea, you're your own boss, effectively. You know what you've got to do and you do it. And you don't have to travel to and from work, you're there … even your food's laid on [laughing].[173]

Adapting to a settled life could be challenging. Women accustomed to managing the household alone could also find it difficult to have a husband permanently at home, particularly under such strained circumstances. Ivy reports that some couples split up and the pressure on relationships is evident in other oral accounts.[174]

Many trawlermen expressed that they missed fishing. Ken's office at the WFA overlooked the lock gates of St Andrew's Dock. For some years he was saddened by the sight of trawlers leaving and returning without

170. Interview with John K.

171. Interview with Christine. Trawler radio officer Mike also uses the term 'itchy feet' to describe the impetus behind his own return to sea from a secure job on an oil platform. Personal Communication, Letter from Mike, July 2013.

172. Interview with Christine.

173. Interview with David.

174. Interview with Ivy. See also T. Creed and J. Hawkins, *The Northern Trawl* (Beverley: Hutton Press, 1986), p.75.

him. It was hard, he says, because 'I missed it. I think it took about four years before I thought to meself, well it's out of your system now and of course by then, the fishing industry was well on its way out.'[175] Amongst those who continued to work at sea, Laurie missed the independence of the trawler skipper and Bob and Bill the daily challenge of fishing. Victor missed that special camaraderie forged by working and living with an extreme seafaring occupation. To Jim, trawlermen were hunters at heart. Even with advanced technology, they hunted for a living and Jim says that is 'in your blood'.[176] A collective buzz of anticipation is evident in many recollections of the trawling years; the thrill of the chase and the expectation of the big catch. Gill, speaking of her husband John, says, 'even when he came home, he always talked about fishing and when you met men [*who were*] fishing, they talked … cos I used to say, well, I'm sick of hearing about how many and how much you got of fish.'[177] Such a buzz is conspicuously absent in narratives of later times. Even for those remaining in trawling, the new practices of international landings and irregular returns to the home port diluted the collective experience. Christine recalls the enjoyment of past settling days, when trawlerman would return to the fish dock to collect their money. She feels that the sense of occasion was lost when crews no longer sailed regularly from Hull. 'They didn't have that rapport,' she reflects, 'y' know, what they used to have when they went down and saw each other. It just went. And they just didn't go to the office much. [*Their money*] was put into the banks then.'[178]

The fishing community had grown accustomed to comings and goings that took place at Hull's fish dock, in many cases just a short distance from their own front doors. The dispersal of the fishing fleet and the move into other seafaring careers brought new travel arrangements. In global fishing, arrival in port did not always mean arriving home. A further journey could be required, by coach, taxi or even aeroplane, before being united with family and friends. The same applied to the outward trip. Even though men had regularly worked hundreds of miles from the city, a starting point other than Hull's fish dock served to emphasise the distance. Gill describes her reaction when her husband John took a job in the Gulf:

175. Interview with Ken.
176. Interview with Jim.
177. Interview with Gill.
178. Interview with Christine.

> oh, I went [*said*], oh going all them miles. And he went [laughing] – well, it's no further than going to, like, Canada, off Newfoundland and all that. But it was, cos I was taking him to a plane. It won't like going away [*when*] I used to take him to the ship, or the bus would pick 'em up when they was mackerelling.[179]

Gill was also alarmed by the length of time that John would be away. Although the traditional 21-day trawling trip offered only a short time at home, it is clear that many preferred this frequency of contact to longer absences. After 1976, there could be little choice, as Gill once again illustrates. 'I said, you can't go. You can't leave me for three months. It's endless,' she recalls. 'So, he said, well, it's work and if I don't work, you won't be able to live to the custom that you have been living.'[180]

With her husband regularly away for longer and her children getting older, Gill decided to go out to work herself; a departure from her former life as the wife of a trawlerman. For the women of the fishing community there were other changes too. With men in new careers or trawling overseas, the coherence of matriarchal support networks began to dissolve. Like camaraderie at sea, such networks had grown from shared experience and shared need. After 1976, although many men were still at sea, diverse new roles replaced old commonalities and rendered former ties redundant. Individual friendships endured, but the broader social links with other fishermen's wives became harder to sustain and arguably less essential for getting by. The weakening ties left a void. Pauline found that her new life as the wife of a merchant seaman could feel more isolating without the support of the female networks that she had established within the fishing industry. Pauline, like Gill, had kept in touch with closer friends, but lost contact with many others. In the past, Pauline had socialised with other fishermen and their wives and they had taken their children out together when their husbands were away. But, she says, 'as soon as the fishing industry went … it just went. It died. It died a death really, as friends.'[181]

The emerging new lifestyles carried compensations. The elimination of the 21-day trip meant more time for home, family and leisure. Some, like Ken, were now working ashore. For Jim, sailing pilot cutters on the Humber, working close to home meant that he could see his family every

179. Interview with Gill.
180. Interview with Gill.
181. Interview with Victor and Pauline.

day.[182] For those working in global waters, longer spells away generally resulted in longer periods of leave. For Gill's husband, longer shore time had occurred too late for him to see more of his own children growing up, but it did bring extra time with his grandchildren. There were some regrets. Gill says, 'He used to say to me, why didn't I have the time like I have now, with my children.'[183] Victor, working in the Merchant Navy, found there were unexpected opportunities to see his family. On occasion, Pauline and their children would be permitted to travel to ports in Spain, Holland, Belgium and France and to stay aboard his ship. It is something that would have been unthinkable in his trawling days.[184]

For those still at sea, longer spells of leave allowed time to plan a holiday, relax and even have hobbies. Michael and Gill's husband John learned to drive; something many trawlermen had deemed low priority when they had just three days ashore. 'He never had time,' Gill explains. 'He'd have a few driving lessons then he'd give it up. He'd say – oh, I'm going for a pint, I can't be going on driving lessons. You didn't have the time for it. But when he went in the Gulf, he learnt to drive.'[185] Having more time and being able to drive brought rewards. As a trawler skipper, Michael says, 'you had like blinkers on'. Michael had followed Hull City football and the racing, but with three days ashore and with stresses at sea, he says that previously there had been little opportunity for outside interests. But with a car and leisure-time, life was different. 'I used to go up North Yorkshire and places,' he says, 'and just park somewhere and just walk all day or go fishing or whatever and it was lovely.'[186]

A further shift for those at home was the reduction of anxiety. Wives, daughters, sisters, all reported living with worry during the hazardous days of Arctic trawling, particularly in the era of the sidewinder fleet. Following the Cod Wars, for those with men working amidst the unstable regimes and precarious politics of global operations, there could be new fears to contend with. Gill, for example, became anxious when John's ship was involved in minesweeping operations during the 1990s Gulf War.[187] For Pauline, however, when Victor left trawling it released her of the fear that he may

182. Interview with Jim.
183. Interview with Gill.
184. Interview with Victor and Pauline.
185. Interview with Gill.
186. Interview with Michael.
187. Interview with Gill.

be lost, which she had carried through her earlier life. Pauline expresses this within a conversation about superstition. Superstition was common amongst many seafarers, but it was particularly prevalent within trawling.[188] By the 1960s, Tunstall reported that superstition was waning, with his respondents attributing it to the older generation.[189] Both Gill and Pauline showed scepticism towards superstition. Nevertheless, they were bound to its practice; sometimes by the older women in the family, sometimes by the fear of regret should a disaster follow a break with tradition.[190] Their narratives hint not only at the strength of the custom, but also at the power of matriarchal networks and cultural dialogues in moulding expected behaviour. When Victor joined the Merchant Navy, Pauline says of superstition, 'I never bothered. It just went.'[191] Pauline places this turnabout in the context of the many big changes that were happening in her life:

> Your life was changing. Your money was changing. Your kids … life was changing … so going to sea in the big ships was never a worry to me. When he was trawlering, I mean … I love the life, I loved the people, I loved him, I loved what he did. But you still always had that big fear with fishing.[192]

For fishermen too, new careers could bring relief from the anxieties of trawling. This was particularly so for trawler skippers. Although some expressed boredom on standby vessels and others lamented the loss of independence, like Laurie in his 'North Sea taxi', there were compensations. Released from the responsibilities and insecurities of landing a lucrative catch, Gill feels that her husband felt less stress. She reflects, 'I look at life, how he was when he was fishing and when he got his tickets and about the catch, trying to get the catch and psychologically what it does to you as

188. A. Gill, *Superstitions: Folk Magic in Hull's Fishing Community* (Beverley: Hutton Press, 1993).

189. Tunstall reports, 'If asked about the question of superstition fishermen usually say vaguely that fishermen used to be superstitious in the old days, or they say: "Ask one of the old men about it".' Tunstall, *The Fishermen*, p.168.

190. Pauline's mother-in-law and Gill's grandmother reinforced superstition. Gill reports breaking taboos. Pauline calls it 'codwash', but stuck to some practices because if something had happened following a breach 'you would've blamed yourself'. Interviews with Jill, Victor and Pauline.

191. Interview with Victor and Pauline.

192. Interview with Victor and Pauline.

well.' Gill goes on to conclude that John was 'a lot more relaxed man in the Gulf'.[193] Michael too describes a calmer working atmosphere:

> It's unbelievable. When I went to Ghana … all that pressure'd gone. I couldn't believe it. Not just me, other people was the same I would think … I can't explain how different it was. How easy you felt in yourself, having a job to go to – that's the main thing – cos fishing has always been casual, where you don't know whether you'd have a job because you had the wrong coloured eyes or the wrong suit on or someone didn't like you … all that was stressful. So, when you got on survey ships it was lovely and you met decent people and you won't worried about fishing or expenses anymore.[194]

Beyond the Cod Wars, much had been lost, but some things had been gained. With hindsight, Michael reflects upon the transition:

> I was skipper for over twelve years, so you get used to it – the good and the bad and you feel happy in yourself at what you're doing. So, I was really disappointed when we got thrown out of Iceland and it all went down the tubes. But after a while, it was like a novelty, meeting all these people on the ships who'd been to university, cos I'd never met people like that … so that was interesting. And then when I went to Ghana and saw sunshine and palm trees, I wished I'd been there 50 years before, it was really nice.[195]

For Michael and others like him, the sting of losing an established occupation and distinctive way of living could be gradually soothed by positive new experiences. For trawlermen, there were emerging new careers: adventurous for some, tedious for others. Some had gained stability and satisfaction, like Michael, who speaks with enthusiasm of his post-trawling activities. But others struggled from contract to contract, between sea and shore, between work and prolonged unemployment. In the home port, the certainty of work in the ancillary and fish trades had also been shattered. Jobs had been lost and where trades had survived, the work available had diminished. In the troubled British economy of the 1970s and 1980s, men

193. Interview with Gill.
194. Interview with Michael.
195. Interview with Michael.

and women from Hull's former fishery joined the ranks of the unemployed and underemployed that seeped from the city's port and manufacturing sectors. Experience, work ethic and a willingness to venture away could secure new positions. But the outcome disturbed the dynamic of Hull's fishing district, and statistics of employment and deprivation in the Hessle Road area suggest that the collapse of trawling had an impact on the next generation.[196] Deckhand Bill reflects. 'I've got two sons,' he says, 'they'd have carried on and other fishermen with sons, they'd have carried on. And if [*there was*] fishing today, all the sons, they'd have carried on. But [*there*] isn't. It's gone. There was work for everybody. You didn't have to go to sea ... there was different jobs. Now there's nothing.'[197]

1983 and the Last Firms Standing

In 1983, the EEC finally released the long-awaited revised CFP, ending a decade of uncertainty. That decade had witnessed transformations in global fishing that had sent the British distant-water trawl industry into turmoil. The absence of clear policy had hampered attempts to restructure.[198] This was particularly so for the fish-catching sector. With limited financial support from the UK government, trawling firms had struggled to accommodate rapid change, whilst grappling with fluctuating quotas and undermined by competition from the better subsidised fleets of other nations.[199] The 1983 CFP brought a framework for adaptation and financial aid for the decommissioning of vessels that were surplus to requirement. For the firms of Hull's fish dock, the package came rather late. The sidewinder fleet had already gone, reduced from 71 vessels in 1971, to five in 1981 and just one – the *Arctic Corsair* – in 1986.[200] In 1983, the freezer fleet stood

196. Loader's analysis of census data shows that between the 1970s and the 1990s unemployment in Hull was above the national average. Her research is inconclusive regarding to what extent unemployment and high deprivation levels reflect the decline of fishing, the wider recession or other local factors. However, she points out that the 1981 statistics do show the highest levels of unemployment in the wards adjacent to the fish dock, and that high levels of unemployment persist in these wards in 1991. Loader, 'Decline of Hull's Fishing Industry'.
197. Interview with Bill.
198. SFIA *Hull Fish Industry in 1983*, p.2.
199. SFIA, *Hull Fish Industry in 1983*, pp.97–101.
200. Thompson, *Hull's Side-Fishing Fleet*, p.94.

at 17 vessels, of which only 11 were active and sea-ready.[201] Economic pressures had seen Hull's once extensive distant-water fleet scrapped, sold and scattered. European aid for decommissioning merely mopped up the remaining surplus.

A view exists that in the face of the post-Cod War challenges, British trawling firms simply walked away. In Hull, the evidence is to the contrary. All firms struggled on, investing effort, energy and depleting finances into diversification. But in the years leading up to the finalisation of the revised CFP, adverse forces determined their fate. By the late 1970s, the smaller company of Newington Trawlers had abandoned distant waters in favour of inshore and North Sea operations.[202] In 1983, Thomas Hamling & Company went into receivership, just short of its centenary.[203] By 1982, Boston Deep Sea Fisheries had lost £3 million over three years and was under pressure from its board to cease fishing operations.[204] In 1983, the company was taken over by the North British Maritime Group and by 1986, their fishing fleet had been discarded.[205] Also in 1986, British United Trawlers sold the last of their freezer ships to foreign owners.[206]

As the British trawl industry moved towards a new era directed by a revised European CFP, only two trawling firms remained on Hull's fish dock. From 1986, J. Marr & Son[207] and Boyd Line were the last remaining fishing companies. With persistence and resourcefulness, both firms had weathered the perfect storm, progressing from traditional cod-focused North Atlantic fishers, to diverse and versatile companies able to work flexibly in a global marine economy. From the mid-1980s, the fallout from a decade of disruption was beginning to settle. With the added security of the CFP, the British fishing industry faced a period of relative stability and modest recovery.[208] For the distant-water sector, there was also the promise of the new British fishery in the South Atlantic. In 1988, *World Fishing* was cautiously speculating that perhaps after all, there was a place for a new

201. SFIA, *Hull Fish Industry in 1983*, p.9.
202. Thompson, *Hull's Side-Fishing Fleet*, p.94.
203. Thompson, *Hull's Side-Fishing Fleet*, bid., p.20.
204. *FN*, 26 March 1982, pp.6–7.
205. Thompson, *Stern Trawling Fleet*, p.20 and p.36.
206. Thompson, *Stern Trawling Fleet*, p.52.
207. In 1986, J. Marr & Son was restructured into two companies: Andrew Marr International, dealing predominantly with fish, and J. Marr, dealing with fishing and survey vessels.
208. *WF*, October 1986, p.30.

generation of big trawlers, equipped to diversify from earlier dependencies upon cod.[209] In that year Marr took delivery of three large freezer vessels. The first, *Thornella*, was the largest trawler to have been built in the UK for over a decade.[210] In Hull, that decade had witnessed a colossal sacrifice to the changing regimes of international fishing. But there were survivors, and for those who were left, in the fish trades and in the decimated catching sector, there was for the time being the chance of a modest future.

209. *WF*, January 1988, p.1.
210. *WF*, April 1988, p.51.

Chapter Six

The Life and Death of Trawlertown

From its earliest days, the fishing industry in Hull had developed firm place-based associations. Localised operations and business structures had given the fishery a strong physical and cultural presence in the portside district of Hessle Road, where trawling had moulded the urban landscape with dock, railway and factory. Economically, fishing provided livelihoods at sea and on shore, employing local men and women who frequently, although not always, originated from Hessle Road. It underpinned the high street shops that traded in the fish dock vicinity, putting money in the pockets of those who lived and worked there. In return the district provided labour for every aspect of its operation. This chapter takes us ashore. Hessle Road is portrayed here as Trawlertown: a fishing heartland, where the comings and goings of the fleet, the celebrations, losses and the daily shoreside grind took place amidst the bricks and mortar of terraced streets running down to the dock.[1] It is where the networks of female support and companionship described by Gill and Pauline were strengthened by daily contact. For concentrated here were the families of fishermen, dockworkers and fish processors who shared and understood the centrality of the fish dock to their lives.

The 1970s brought disturbance to Trawlertown. The disruptive technology of the freezer trawler was diluting the collective experience and culture of the 21-day trip. At the same time, Hessle Road residents were facing

1. Elements of this chapter have previously been published in the chapter J. Byrne, 'Hull, Fishing and the Life and Death of Trawlertown: Living the Spaces of a Trawling Port-City', in B. Beaven, K. Bell and R. James (eds), *Port Towns and Urban Cultures: International Histories of the Waterfront, c.1700–2000* (London: Palgrave Macmillan, 2016). The term Trawlertown is associated with the folk CD, J. Conolly and P. Sumner, *Trawlertown: The Singing of the Fishing* (Fellside, 2000) [CD]. Predating this is the local history video *Trawlertowns – Hull and Grimsby* (Forest Edge, 1993) [VHS].

the breakup of their tight-knit community as post-war housing clearances gathered momentum. Amid this transition came the Cod Wars and the ensuing decline of trawling. As Hull's distant-water fishery suffered a decade of upheaval, the impact was felt in the dockside district. The search for new opportunities took ships and men away from the city and brought new suppliers to the quay. The social, economic, spatial and cultural connectivity that had defined Hull's fishery for almost a century was interrupted, and a maritime community was divorced from the industry that had given it life.

Coming into Port

A port is more than its docklands. Nevertheless, historians, geographers and sociologists alike have often maintained a narrow focus on the 'immediate connections between land and water'.[2] Port historian Sarah Palmer observes an 'inclination to look out to sea rather than inland, or to allow the dock wall to define the limits of investigation'.[3] Historians have conceptualised port–city relationships based upon multiple social, economic and spatial linkages. Palmer has revealed associations relating to transhipment, labour, mercantile services and processing industries. These economic relationships were detectable spatially, making the port city a recognisable physical entity.[4] At the same time, the port city responded to the social needs of seafaring, enabling those engaged at sea or in a maritime occupation on land to reside in a neighbourhood that understood and shared their experience.[5] The idea of a defined maritime quarter responding to the needs of the seafarer ashore is encapsulated in Hugill's *Sailortown*.[6] In Hugill's generic sailortown, the transient demands of the fluid crews of deep-sea sailing ships were met in a dockside district of pubs, brothels, pawn shops, marine stores, tattoo parlours, boarding houses, music halls and tailors who would lend on advance notes.[7] For Hugill, it was 'a world in, but not of, that of the landsman'.[8] Hugill's work is a popular and at times nostalgic depiction of

2. S. Palmer, 'Ports', in M. Daunton (ed.), *The Urban History of Britain* (Cambridge: Cambridge University Press, 2000), p.133.
3. Palmer, 'Ports', p.133.
4. Palmer, 'Ports', pp.138–139.
5. Palmer, 'Ports', pp.148–150.
6. S. Hugill, *Sailortown*, 2nd edn (London: Routledge & Kegan Paul, 1967).
7. Hugill, *Sailortown*, pp.74–87.
8. Hugill, *Sailortown*, p.xviii.

the port city district. Nevertheless, the idea of sailortown has been adapted by Daunton, examining Cardiff's port district in the nineteenth century, by Hilling, considering Cardiff in decline and by Palmer, examining the varied character of port–city relationships.[9]

More recently, Moon's study of the nineteenth-century sailortown within Portsmouth has revealed a more settled relationship between those at sea and those on land.[10] Although a century apart, Moon's findings resonate with Hull's late twentieth-century fishing district, for like Hull trawlermen, the seafarers in Moon's study were not passing strangers, but were married and integrated into a supportive community through occupational, familial and local ties. Moon has identified a network of residential 'sailorhoods', where a sea-facing culture produced distinct attitudes and practices among residents and seafarers alike. Although spatially defined and demographically distinct, Moon's 'sailorhood' is not the homogenous 'other space' of Hugill's *Sailortown*. Instead, it is entwined with the wider urban communities and districts that coexisted within Portsmouth.[11]

Moon's work is situated within a growing cultural approach to port studies that has seen topics as diverse as cinema going and the supernatural examined as indicators and mediators of the bond between land and sea.[12] In an era following the decline and remodelling of many waterfront districts, memory too has been deployed as a means to express the changing spatial, societal and cultural dynamic of the port city.[13] This chapter builds upon such cultural approaches, engaging memory to articulate the port city as a network of meaningful lived spaces and emotional affiliations. The connection between people and place is explored through

9. M. Daunton, 'Jack Ashore: Seamen in Cardiff before 1914', *Welsh History Review*, 9 (1979), 176–203; D. Hilling, 'Socio-economic Change in the Maritime Quarter: The Demise of Sailortown', in D. Pinder and S. Hussain (eds), *Revitalising the Waterfront* (London: Belhaven Press, 1988); Palmer, 'Ports'.

10. L. Moon, 'Sailorhoods: Sailortown and Sailors in the port of Portsmouth circa 1850–1900', Unpublished PhD Thesis, University of Portsmouth, 2015.

11. Moon, 'Sailorhoods'.

12. R. James, 'Cinema-going in a Port Town 1914–1951: Film Booking Patterns at the Queens Cinema', *Urban History*, 40:2 (2013); K. Bell, 'Port-city Afterlives: The Cultural Function of Ghosts in Nineteenth-century Portsmouth', unpublished paper presented at Port City Lives Conference, Liverpool University, 29–30 June 2012.

13. A. Mah, 'Memory, Uncertainty and Industrial Ruination: Walker Riverside, Newcastle upon Tyne', *International Journal of Urban and Regional Research*, 34:2 (2010); L. Balderstone, G. J. Milne and R. Mulhearn, 'Memory and Place on the Liverpool Waterfront in the Mid-Twentieth Century', *Urban History*, 41:3 (2014).

narratives of everyday engagements, habitual movements and situated social interaction. Recollections of embodied encounters evoke a seafaring world shaped by practice, smell, sound and sensation. They foreground a sense of place that is deeply rooted in the human experience, revealing meanings and attachments, which despite originating in the past, resonate through memory into the present.

The centrality of embodied lived experience in the creation of place is encompassed within Ingold's concept of the taskscape. In 1993, social anthropologist Ingold conceived the taskscape as a means of understanding place as the located ensemble of collective tasks, sensations, sounds and encounters performed in the process of communal living. The taskscape is a network of what Ingold refers to as the activities of dwelling: the everyday tasks that we undertake in our everyday lives. Activities are ongoing and temporal. Temporality is presented as the passage of time as experienced from within the event, and as social time spent in communal activity.[14] To become a taskscape, action must be collective and take place as part of a shared and ongoing lifeworld. Ingold writes, 'It is to the entire ensemble of tasks, in their mutual interlocking, that I refer by the concept of *taskscape*. Just as the landscape is an array of related features, so – by analogy – the taskscape is an array of related activities.'[15] Importantly, Ingold relates these everyday temporal processes to the simultaneous creation of physical landscape. The taskscape and the landscape arise together within the same flow of activity, just as a man's gait and his footprint arise in the movement of his walking.[16] The taskscape offers a theoretical perspective on how the movement, processes and connectivity of ongoing everyday life can, over time, physically mould the landscape, as well as imbue it with collective meaning.[17]

Ingold presents the taskscape as continually unfolding, with the landscape never reaching completion so long as people are collectively engaged within it. Even when activity ends, a legacy remains. Activity is physically expressed within our surroundings, and the landscape can therefore be regarded and retained as a 'collapsed act'.[18] Ingold writes, 'thanks to their solidity, features

14. Ingold, 'Temporality of the Landscape', pp.159–160.
15. Ingold, 'Temporality of the Landscape', p.158
16. Ingold, 'Temporality of the Landscape', p.162.
17. In an extension of Ingold's concept, Benediktsson and Waage have also examined the multiple taskscapes evident in the contained yet fluid world of the fishing trawler at sea. K. Benediktsson and E. Waage, 'Taskscapes at Sea: Trawler Fishing and the Experience of Fluid Nature', *Environment, Space, Place*, 7:2 (2015).
18. G. Mead, 'The Process of Mind in Nature' (1977[1938]), quoted in Ingold, 'Temporality

of the landscape remain available for inspection long after the movement that gave rise to them has ceased.'[19] This is significant, for it means that the physical relicts of taskscapes occurring in the past may survive to gain new collective meaning in the present. Ingold's phenomenological, sensual, yet simultaneously physical rendering of place can be adapted to explore the process and experience of change. In this chapter, through oral history narratives, the taskscape of Hull's fishing district is made visible as *Trawlertown:* a recognised and lived entity embedded within the flows of the wider city. When fishing activities cease, and the social, economic and cultural connectivity that created its maritime spaces are consigned to the past, remaining in their place is memory.

Stepping into Trawlertown

Oral history provides an ideal vehicle for a visit to Trawlertown. As a phenomenological encounter, narrators can invoke its existence because they lived it and lived in it. Respondents recall an exceptional neighbourhood where the taskscapes of fishing could coexist with other expressions of community. Trawlertown cannot be divorced from the broader composition of Hessle Road, just as Hessle Road cannot be ascribed exclusively to fishing. Lives and lifeworlds intertwined. But with a nuanced eye, it is possible to distinguish Trawlertown amid the bustle of a working-class district. The practices and culture of distant-water trawling – the concentration of shoreside activity, the casual employment, fluctuating pay, short time ashore and the three-day millionaires – created an environment suggestive of a latter-day sailortown. On Hessle Road until the 1970s, the fishermen's pubs, clubs, pawnshops, chandlers, backhanders and loan systems, even the fishermen's outfitters and tailors, draw easy comparison with Hugill's popular depiction or with the sailortown districts examined by Daunton, Hilling or Palmer.[20] Yet here the comparison ends, for, like the seafarers in Moon's study, Hull's fishermen were not just returning to port. They were returning to Trawlertown, their home community and to wives, mothers and girlfriends, family, children and shore-dwelling friends.

In his analysis, Horobin explains the circumstances that had created

of the Landscape', p.162.

19. Ingold, 'Temporality of the Landscape', p.162.

20. Daunton, 'Jack Ashore'; Hilling, 'Demise of Sailortown'; Palmer, 'Ports'.

the bond between the fish dock and the adjacent district of Hessle Road. He describes trawling as an industry that was 'essentially different from any other and which by its nature affects the whole social life of its workers'.[21] In the 1950s, fishermen constituted around 11 per cent of Hessle Road's male workforce, yet Horobin names them as the key group in an industry that dominated the neighbourhood. As a result, Hessle Road became an area where 'practically the whole of the population … including shop keepers, were "geared" to the rhythm of fishing'.[22] Here, shops were reported to vary their prices in accordance with the state of the market for fish. Such interdependency was characteristic of single-industry settlements, such as mining villages, but Horobin sees this as less common within districts of cities the size of Hull. He suggests that dependency upon the industry fostered high levels of community interdependence, which became particularly apparent in times of hardship and disaster. 'When a vessel is lost,' says Horobin, 'the whole community "draws together", as it were, for the bell tolls for everyone, and everyone knows it.'[23]

The position of Hessle Road at the heart of the fishing community is taken for granted by narrators, regardless of whether they themselves lived there. A visit to Trawlertown was a phenomenological encounter with the port city. The fishing district had economic and material expression, with fish dock, smoke house, railway and cold store immediately recognisable as the fabric of a trawling port. But the essence of *Trawlertown* lay in the collective encounters of those who lived and worked there. Trawlertown was a lifeworld that was walked, smelt, heard, felt and touched. It was lived space, embodied place, and although it could be observed from the outside, it could only be fully understood from within.

Hessle Road: 'In Full Swing'

Narrators in this chapter have a mix of associations with Hessle Road. There are 'Hessle Roaders' born and bred, some, but not the majority, living there still.[24] A small number of seagoing narrators had joined the fishing industry from other parts of the city, yet they still narrate a strong affinity with the fishing district. Most other seafarers originated from Hessle Road,

21. Horobin, 'Community and Occupation', p.348.
22. Horobin, 'Community and Occupation', p.348.
23. Horobin, 'Community and Occupation', p.348.
24. Correct as of 2014.

reinforcing the notion of trawling as a neighbourhood tradition. For some, Hessle Road was a workplace only, the district that they travelled to, worked within and left. These narrators were generally, but not exclusively, the more senior members of the industry: the trawler owners, firm directors, technical professionals and some fish merchants. Trawlertown was composed of diverse connections and fluctuating encounters. Binding them together were the relentless rhythms of fishing.

Narrators were invited to describe Hessle Road in the 1960s, when the fishing industry was 'in full swing'. Responses may at times blur the timeline, and convey instead a post-war heyday, where memories of the 1950s and 1960s combine. In many respects, memories of Hessle Road in this era resonate with recollections of other working-class communities at this time.[25] Klein defines a traditional working-class community as one of close-knit social networks produced by 'the concentration of people in the same or similar occupations in the same local area, little migration into or out of the area, local intermarriages, continuity of social relationships, opportunities for relatives and friends to help one another, little opportunity for physical mobility, little opportunity for social mobility'.[26] Just as Roberts witnessed in the working-class districts of Lancashire, narratives of Hessle Road portray a community that looked after itself through mutual assistance, was self-policing and forged social networks amidst local streets, shops and pubs. Lily echoes others in describing the security of belonging to a place where you were known. She reflects:

> it was like a big family. I mean, no … that isn't fair, it wasn't like a big family, but … there were a lot of people you knew. Some you only knew by sight to say hello to, some you knew really well and you had proper neighbours … and if anything happened, they were really kind. I've known people and they've took kiddies in or they've made meals and just taken them across.[27]

There were challenges and constraints to communal living. As a child, Lily says, 'you could never get away with anything … because if your mother

25. J. Klein, *Samples from English Cultures* (London: Routledge & Kegan Paul, 1965); E. Roberts, *Women and Families: An Oral History, 1940–1970* (Oxford: Blackwell, 1995).
26. Klein, *English Cultures*, p.130.
27. Interview with Lily.

didn't see you do it, somebody did'.[28] It was not only children who were policed by the community. Staff working in David's fish merchant business came from the Hessle Road area and David recalls one older man who took on the role of community champion. 'He wasn't a violent man,' David reflects, 'but he would sort everything out that went on in [*his*] terrace.'[29] Community could offer comfort and belonging, but as Lillian Bilocca and other leaders of the women's campaign for trawler safety discovered, breaking its unwritten codes could result in alienation or retribution.[30]

Accounts of Hessle Road often convey the endless movement and multitudes of a working-class thoroughfare. Gill recalls the volume of life: 'It was full of people,' she says, 'Hessle Road was full of people.'[31] Ivy describes it as 'a very, very busy place … it used to take you hours to do your shopping, because you knew everybody.'[32] Reminiscent of Seamon's 'place ballets', the surge of daily encounters forged an 'insider's' knowledge and a sense of belonging that circulated around the terraces, amid the children playing in the streets and the women donkey stoning their doorsteps.[33] It congregated in the many pubs that could be used to navigate the district, and convened within the local shops. Almost without fail, narrators refer to the myriad of small retailers that lined the high street – the little epicentres of communal life. Mark, a lifelong resident of Hessle Road, reflects upon the extent of connectivity: 'there was loads of little shops … you could go into a shop and nearly everybody knew somebody who you knew.'[34] Roberts and Klein both suggest that gossip in working-class communities helped to maintain neighbourhood solidarity, regulating social behaviour, fostering belonging and even alerting the community to those in need.[35] The shop stood at the heart of the gossip exchange.[36] In their domestic role, women in particular might visit their local retailer several times a day and the shop as a place for gossip is recalled by

28. Interview with Lily.
29. Interview with David.
30. Despite substantial support from within the community, the women leading the campaign at times faced criticism, ostricisation and even open aggression from men and women who were unhappy with the way that they represented the trawling community, or who resented their interference in a man's world.
31. Interview with Gill.
32. Interview with Ivy.
33. Interview with Ivy.
34. Interview with Mark (name changed), March 2013.
35. Roberts, *Women and Families*, pp.207–211; Klein, *English Cultures*, p.128.
36. Roberts, *Women and Families*, p.210.

Mark: 'There was a lot of backchat going on and chit chat,' he says. 'Did you hear about so-and-so having a baby, sort o' thing. And people knew each other that way, y' know, there was always a bit of banter going about.'[37] Narrators offer a string of familiar stores – Maypole, Meadow, Home and Colonial, Winkles, Fletchers, Boyes, Clothing House. On Hessle Road, Thomas claims that you could buy 'anything from a safety pin to a Rolls Royce'.[38] For Lily it was a little community of its own. Everything that you needed was available and there was no pressing need to go elsewhere.[39]

In the 1950s, Hessle Road was reported to Horobin as 'a world in itself'.[40] My narrators also remember Hessle Road as a self-contained world, which functioned in a wider city but could nevertheless provide for its own. They conjure a self-sustaining 'otherness' that outsiders might perceive as a closed world. Rick was a musician, and an 'outsider' to both Hessle Road and to Hull. He found it impossible to break into the music circuits to perform in Hessle Road clubs. This had not been a problem elsewhere in Hull and he attributes the invisible barrier to the close-knit insularity of this part of the city.[41] Some narrators report that fishermen were looked down on by other city residents.[42] Tunstall suggests that even within the Hessle Road area itself, the streets closest to the fish dock could be viewed with disdain by neighbours located further away.[43]

Outside attitudes, coupled with internal self-sufficiency, served to perpetuate a sense of being different. Narrators describe Hessle Road in the 1960s in terms of the exceptional – as 'something very special', 'a wonderful, wonderful road', 'a breed of their own'.[44] With clarity they connect this distinct sense of place to trawling. For Hessle Road fishmonger Billy, 'the atmosphere of Hessle Road was from nowhere else. There's nowhere else like it. Absolutely, it could only be that [*fishing*].'[45] Ivy holds a similar view:

Ivy That's what Hessle Road was built for. For the fishing industry. Nothing else.

37. Interview with Mark.
38. Interview with Thomas.
39. Interview with Lily.
40. Horobin, 'Community and Occupation', p.243.
41. Interview with Rick (name changed), July 2013.
42. Interviews with Ben and Philip.
43. Tunstall, *The Fishermen*, p.93.
44. Interviews with Ken, Victor and Pauline and Billy.
45. Interview with Billy.

Jo (Researcher)	And could you see that when you were living down there?
Ivy:	Oh yes. I didn't know nothing else apart from fishing. I didn't even know there was a Metal Box [*local factory*] 'til I left school, because that was Gypsyville [*adjacent area*] and I never left from Hessle Road 'til I was 15. I didn't even know there was a town [*city centre*].[46]

Interviews convey the dominance of Hull's distant-water fishery over the Hessle Road district. The taskscape is never far away. Trawlertown comes into being in recollections of the continuous day-and-night movements of dockworkers, fish bobbers, merchants and processors, with the trains and fish vans and with the sound of a ship's hooter. Jim refers to the clatter of 'hundreds' of fish bobbers walking to work in the very early morning in iron-shod clogs 'like horses' hooves'.[47] Alan J. recalls the perpetual threading of people to and from the fish dock. 'You had to walk all the way down West Dock Avenue, under the tunnel, under the railway lines and onto the dock,' he explains, 'and I mean there were always masses of people. You never walked down West Dock Avenue on your own [laughing]. There was always people backwards and forwards all the time.'[48]

Many concur that a visitor to the area would have recognised a connection with fish. There were the filleting factories and distinctive lines of slender chimneys belonging to the kipper houses. There were women wearing the 'wellies' and 'turban' headgear of the fish processor.[49] And there was the odour. 'We did have our own fish smell' admits Ken.[50] Not the seaside smell of a fresh landing, but the heavy, dull tang of fish meal, cod farm and smoke house.

If the area beat to the rhythm of trawling, then the comings and goings of the fleet set the tempo. Hessle Road shops included outfitters 'open to suit all tides' and shops offering credit until dad came home from sea with his wages.[51] Fishermen feature prominently in narratives

46. Interview with Ivy.
47. Interview with Jim.
48. Interview with Alan J.
49. Interview with Lily.
50. Interview with Ken.
51. Gill, *Hessle Road*, p.29.

Figure 9. Visible in their wellies and headgear, fish factory workers wait at a Hessle Road bus stop in 1979 (Copyright Alec Gill).

Figure 10. Taxi offices, Subway Street, Hessle Road, 1979, within easy reach of the fish dock (Copyright Alec Gill).

of Trawlertown. Trawlermen were known for wearing handmade suits in the latest and sometimes extravagant fashions, making them instantly recognisable. 'They had their own sort of tailoring style,' says Philip. 'Double breasted suits, which were tightly fitted round the hips, the bell-bottom trousers. In outrageous colours as well some of them.'[52] Prior to a husband's homecoming, fishermen's wives would bustle to retrieve suits that had been pawned to enhance the family budget, and once home they carried work gear to the wash house, ready for the return to sea. Many observed a visible generosity and easy spending, along with the frenetic movement between home, dock, pub, the homes of friends and relations, the tailor, the bank, the barber, as three weeks of living was compressed into three days. Taxis darted here and there, for fishermen did not waste time walking: 'it was a taxi,' says Ivy, 'even though it was only next street.'[53]

Ken's narrative shows how the interactions of fishermen could infuse everyday physical spaces with specific collective meaning:

> there was your own tailors … We had our special shoe shops, they all knew the fishermen more or less by names. We had our own barber that would give you a dram o' rum while you were sat in his barber's chair. You'd all be sat talking while you're waiting for your hair cut – about how much you'd made and how much you landed – and so all the way along the length of Hessle Road, there was stories of the fishermen.[54]

In such places, fishermen came together to reconnect back to land, and at the heart of the process were the pubs. Narrators list the Dee Street Club, St Andrew's, Halfway, Wassand Arms and of course the Star and Garter, the fishermen's pub, known to 'insiders' as Rayner's. Margaret was a barmaid at Dee Street Club. She recreates the regular flow of new arrivals, many of whom she knew by name:

> Right, rack 'em up again, Margaret, from the back of the room. All right. Same round? Same round … Then there'd be another crew come

52. Interview with Philip.
53. Interview with Ivy.
54. Interview with Ken.

> in an'all. I'd think, Jesus. I'd be one end o' there, to there. All day long, but I loved it.[55]

Pubs were more than just places to drink, says Jim, they were places to meet your friends. Tunstall observed that certain clubs on Hessle Road were regarded by fishermen as their 'exclusive preserve'.[56] Relatives would know which to frequent in order to meet with a newly arrived crew. Whilst some harboured the hope of a generous round, most came to welcome and catch up. With only three days ashore, pubs were not merely for leisure, they were important arenas to maintain relationships.

Fish Dock

At the end of West Dock Avenue, through a tunnel under the railway lines of the LNER,[57] stood St Andrew's Fish Dock: the very hub of Trawlertown. Here, in the transitional space between land and sea, the actions, encounters and fabric of Hull's fishing taskscape were generated at their most intense. The fish dock was another world; a geographic entity that was seldom seen by those without reason to visit. Alan J. remembers his first visit to the dock in 1960. 'I knew we had a big fish industry,' he says, 'but I couldn't envisage what it was like. But when I went down there, it just amazed me, because there were just hundreds and hundreds of people, doing all sorts of things … when I saw it, I thought – this is huge.'[58] Alan's job with a trawling firm pulled him into the heart of the fishing taskscape and its endless streams of work and people. Narrators draw upon common terms to recount the animation of the working dock; 'very, very busy', 'hustle and bustle all the time', 'a beehive', 'like ants'. They convey a space of anticipation. Ron describes the buzz and noise of the Dutch-style fish auction, with its jostling buyers and salesmen standing on large aluminium fish containers known as kits. The thrill of the big catch resonated through the trawling community and Ron extends that excitement to the dock. 'There was an air of expectancy,' he says, 'and there [*was*] a vibrancy about it all the time'.[59]

55. Interview with Margaret.
56. Tunstall, *The Fishermen*, p.138.
57. London and North Eastern Railway.
58. Interview with Alan J.
59. Interview with Ron.

Descriptions of the fish dock invoke a powerful blend of movement, sound, sights, smell and feelings. Jim describes walking through the seeming disarray:

> It was dangerous walking … You'd have clogs, because there was water and fish slime and ice, melting ice floating … you would have the bobbers landing the fish and at the same time they were bringing up the fish room boards … board scrubbers … there would be shore-riggers splicing wires, there'd be electricians, the Marconi man would be going round all the wireless and the electronics … The cod liver oil boat would be alongside, pumping out … organised chaos – everybody knew what they were doing.[60]

Alan J. remembers the odour. 'I could hardly abide the smell,' he says. 'I mean, the smell in some of the places was awful.'[61] Ben recalls the familiar smells of rope and nets.[62] Michael relays an aural memory. 'When you're on dock,' he says, 'you'd hear people riveting all day – bang, bang, bang, bang, bang – and all the fitters and labourers and it was just endless.'[63] Alan H. recreates an entire soundscape that progressed throughout the day:

> In the earliest hours of the morning the bobbers would clatter down to the dock … in their clogs … there would be noises … of clashes of boards from the ships being thrown onto the quayside and aluminium kits being bounced and people yelling and shouting and the squeal of winches and such like. Then you'd have the auction itself with all its noise and then the frenzy of the fish coming off the dock in trucks and trailers … Peace reigned in the afternoon.[64]

Beyond the spectacle, the dock could be an uncomfortable place to work. Billy, a fishmonger and former barrow boy, describes the dock in the 1960s as 'a bit of a dump'. 'In a wintertime, they'd put bags over their shoulders and put kits up to keep the bloody wind off,' he says. 'On the front there, it was

60. Interview with Jim.
61. Interview with Alan J.
62. Interview with Ben.
63. Interview with Michael.
64. Interview with Alan H.

Figure 11. A forbidden childhood playground: children hang out among fish kits and boxes, 1980 (Copyright Alec Gill).

like pneumonia corner.'[65] As a fish merchant, David also remembers the dock as wide open to the elements. David rented a small office known as a 'stand' along with an adjacent area of the fish quay. Until the vessels finished landing and the fish auction had been cleared, David's workspace was limited.[66]

Some recollections of the dock extend beyond its working role. Ivy and Margaret both remember weekend family strolls to enjoy the sight of ships and buildings.[67] Gill recalls a forbidden childhood playground. However,

65. Interview with Billy.
66. Interview with David.
67. Interviews with Ivy and Margaret.

for fishermen and their families, St Andrew's Fish Dock gained particular significance as the place of arrivals and departures. From distant seas, the dock was the gateway back to Trawlertown and the subsequent point of separation. In the masculine world of the fish dock, superstition dictated that women should not wave off a departing ship. But many families would be there for its return. Margaret recalls meeting her brother and other friends as they arrived into port. 'It was lovely,' she says. 'It made you feel real warm. And glad to see them come back.'[68] For trawlermen, the dock was their first and last contact with land. It assumed territorial significance. 'A dock is a dock,' asserts Ken, 'it's just a place where you take your ship. But St Andrew's Dock wasn't just a dock, it was *the* dock, y' know … it's ours.'[69]

From a mix of affective engagements, shared experience and collective activity, Trawlertown came into existence. Trawlertown was an ongoing lived encounter with the rhythms and spaces of Hull's distant-water fishery, embedded within the West Hull urban landscape. On Hessle Road, other work communities lived side by side with Trawlertown; it was part of their neighbourhood, although they themselves might not engage with it. There were those who breezed in and out, as they laboured in its workforce and drank in its pubs before returning to homes elsewhere. But for many Trawlertown was their primary world. They lived and worked in its midst, unequivocally absorbed within its sounds, smells, sensations and connections. If as some suggest, it is through such encounters that place attachment is forged, then we might expect the exceptional interactions of Trawlertown and its located systems of occupational support to have cemented bonds that would prove difficult to break.

Demolition and Development

From as early as the 1960s, as daily life continued, a series of physical developments were emerging to remould the spaces of Trawlertown. Hull's post-war housing strategy sought to improve living conditions in areas of the city where accommodation was deemed substandard. It included swathes of Hessle Road. By the early 1970s, the fish dock too was under scrutiny as the expanding fleet of freezer trawlers outgrew their shoreside facilities. By 1972, plans were underway to redevelop Albert and William

68. Interview with Margaret.
69. Interview with Ken.

Wright Docks as a new base for the fishery. After more than ninety years, Hull's fishing fleet, its trades and its long-serving community, were set to abandon their established home.

Moving On and Moving Out

> Longhill isn't so alive as Hessle Road. But there's more sunlight and fresh air.
> Uncle Bernard said he didn't think I'd settle up at Longhill and laughed when I told him 'You even miss people you don't like.'
> Annie Eliza Courtney: extracts from Tom Courtney, *Dear Tom: Letters from Home*

After the Second World War, the Hull Corporation faced the challenge of rebuilding the badly damaged fabric of the city. Alongside rehousing the 152,000 people who had been rendered homeless by bombing was the need to continue with area-based improvements that had commenced before the war. Post-war development proposals known as the Abercrombie Plan aimed to deliver a 'proper healthy and amenable state of living' for the city's inhabitants.[70] To achieve this, new local authority neighbourhoods were created to the east of the city at Bilton Grange (1955), Longhill (1958) and Greatfield (1960).[71] A fourth, known as the Thornton Estate, was created in the years up to 1970, in an area between Anlaby Road and Hessle Road, close to the city centre. Further estates developed north of the city at Orchard Park (1963) and Bransholme (1966) and to the east at Ings Road (1963).[72] The new estates would meet the demands of a rising population, whilst dealing with the removal or reconfiguration of substandard areas. The process was referred to as slum clearance and prior to the war had focused on the crowded and insanitary conditions within central Hull. By the late 1950s, Tunstall was observing a shift of attention as the inner-city problems were resolved. He wrote:

> The term 'slum' is now being applied to districts slightly further out. Hessle Road, which was mostly built up between 1890 and 1914 is such

70. D. Neave and S. Neave, *Hull*, Pevsner Architectural Guides (London: Yale University Press, 2010), p.33.
71. Neave, *Hull*, p.33.
72. Neave, *Hull*, pp.33–34.

> a district today, and most of the area to the south side of the road is scheduled for demolition. Many houses have already been pulled down and the threat of imminent demolition hangs over the rest.[73]

As Tunstall was writing, new estates were going up and parts of Hessle Road were coming down. On Hessle Road the process took place over a protracted period, beginning in the late 1950s with the creation of the Thornton Estate and continuing into the 1980s. Alec Gill, who photographed the area between 1974 and 1987, likened certain streets to a battleground. Demolitions had been completed, but national politics after 1979 had delayed the later phases of rebuilding.[74] Indeed, the protraction was such that in the late 1980s new neighbourhoods were planned for the Bean Street area of Hessle Road to re-accommodate tenants from the failing 1960s maisonettes, built on the Thornton Estate during the earliest phases of development.[75] Whilst some were able to benefit from localised rehousing, for many more, particularly in the earlier years, a new house meant a shift from Hessle Road to the outskirts of the city.

The remodelling of Hessle Road impacted upon all its residents. Demolition crews brought large machines and fastened chains around full terraces. 'They just hove on these chains,' Ken recalls, 'and all the houses just collapsed in one big heap. And for people to see that happening must have been devastating.'[76] Conditions may have been substandard, but the demolitions brought a sense of loss. Ivy and Rose both narrate attachment to the very bricks that now lay in collapsed piles. Ivy's past was quite literally etched into the bricks of Manchester Street tram shop – 'I.P. loves L.H.' – the remnants of a teenage romance.[77] Ivy tried to recover the inscribed brick when the tram shop was demolished, but she could not find it. Rose too sought a keepsake following the demolition of her old school:

> Well, we were passing by and we saw the men pulling it down and I said, that's my school, what are they doing to it? So [*we*] stopped the car

73. Tunstall, *The Fishermen*, p.86.
74. Gill, *Hessle Road*, p.111.
75. Gill, *Hessle Road*, p.113.
76. Interview with Ken.
77. Interview with Ivy.

Figure 12. The blitz-like demolitions at Carlton Street, Hessle Road, captured in 1977 (Copyright Alec Gill).

> and I got out and went round and there was this man with bricks and I said, you're pulling my school down. He said, yes. I said, Villa Place. He said, yes. I said, can I have a brick? He said, you can have all the bricks. I said, no thank you, I'd just like one. So, he gave me one brick and I've got it, it's in the shed.[78]

The restructuring had an impact on the form and function of the district. Narrators report the loss of familiar shops and pubs; the little heartlands of daily encounters and gossip. Emptying streets brought challenges to local retailers. Billy remembers the impact on his fishmonger's shop. 'We had no houses over there and no houses at the back, he says, 'but we kept going … Nothing at the back, nothing at the front, but I used to go selling in me van.'[79] Billy had the chance to go mobile, but others did not. As business dwindled, Jim recalls the loss of well-known names. 'The shutters were going up,' he says. 'Shops that you'd used for years … all of a sudden, one

78. Interview with Jack and Rose, February 2013.
79. Interview with Billy.

by one, they'd be closing.'[80] In residential streets, the process of transition could disorder the comfort of home. In the late 1970s, Mark and his mother were the last remaining residents on Beecroft Street, prior to demolition. Mark recalls a disturbing experience:

> There'd been vandalism going on and various people had been setting fire to the houses that was left in that area, so it got a bit scary at times ... There was no telephones either because they'd taken the poles down. And one particular day, me mother said, I can smell burning and when I went out somebody had actually set fire to the house two door away. So, I had to run across St Georges Road and borrow somebody's phone to ring the fire brigade.[81]

Among Hessle Road's disrupted residents were those connected to the fishery. In 1957, Horobin had urged city planners not to disregard the special circumstances of the trawling community when planning for redevelopment. He argued that fishermen benefited from living close to the fish dock, to pick up wages, sign onto ships and minimise time spent travelling during their short spells ashore. The need was even greater for the fish landers or bobbers, who must reach the fish dock at night without the availability of public transport. In 1955, 72 per cent of bobbers against 57 per cent of fishermen lived within one mile of the dock.[82] In terms of recruitment, Horobin recognised trawling as a neighbourhood tradition and advised both industry and planners to consider the consequences of removing trawling families from the area. His most ardent plea, however, related to the social dynamics of fishing. Just as Gill had observed of her husband and his friends, Horobin noted that far from wanting to escape from work ashore, fishermen, like miners, 'talk incessantly about their work'.[83] He saw this as a need for peer group understanding in an occupation unlike any other. The contemporary aspiration to build more diverse or 'balanced' neighbourhoods, he said, ran contrary to this requirement. With this in mind, Horobin believed that the occupational structure of the area should be a key consideration when planning for the future. He concludes, 'We are constantly hearing in Hull about the "clannishness" of fishermen.

80. Interview with Jim.
81. Interview with Mark.
82. Horobin, 'Community and Occupation', p.348.
83. Horobin, 'Community and Occupation', p.348.

Would the balanced neighbourhood break down this clannishness? More importantly, do we want to break it down? Surely not.'[84]

In planning documents for this period, however, the presence of a trawling community is hard to detect.[85] A paper considering the development of the Thornton Estate, albeit closer to the city centre than to the fish dock, does not mention trawling.[86] Occupational issues are discussed in simple terms of travel to work. Residents were asked if they would prefer to live in a flat within the vicinity or have a house and travel a greater distance to their workplace. Their response was 'neither'.[87] Closer to the fish dock, the earliest demolitions focused on the southernmost streets where fishermen and dockworkers lived in greatest concentrations.[88] Gill lived in Flinton Street when first married, and confirms: 'I would think half of that street was fishermen's wives or bobbing … nearly everybody had something to do with the sea, or their family had something to do with the sea.'[89] Several narrators who had lived in this locality had experienced being rehoused. Their stories contain a mix of enthusiasm and regret. Ivy, who relocated to Bransholme via a temporary stay on Beverley Road, was looking forward to raising her children in a greener environment. She occasionally felt homesick and went back to Hessle Road once a week, but she was determined to stay in her new 'castle'. 'Oh, it was going into the country,' she recalls, laughing, 'the smell was different. You could go back if you wanted to wherever you came from. But I thought, no, my children's gonna grow up in the country. It isn't really the country … but it was for us.'[90] John K. too looked forward to moving to Orchard Park and found the move to be 'great'.[91] Lily reports that her cousin, a mother of two, was happy to move away to a new house

84. Horobin, 'Community and Occupation', p.353.

85. HCA Ref: HL711HE. City and County of Kingston upon Hull, Redevelopment of Decayed Area; with particular reference to Residential Area 17, H. E. Alton, May 1960. The documents comprised a limited collection of development and consultation papers relating to Hessle Road district and local plans in the 1970s and 1980.

86. HL711HE. Decayed Area.

87. HL711HE, Decayed Area, p.5.

88. Tunstall, *The Fishermen*, p.93. Tunstall states that the homes of fishermen were concentrated on the south side of Hessle Road, thinning out quickly to the north. He draws upon figures from Horobin's 1955 thesis, which shows 29.9 per cent of all fishermen living within a half mile of the fish dock.

89. Interview with Gill.

90. Interview with Ivy.

91. Interview with John K.

with a bathroom. But for Lily's grandmother, the situation was less positive. Although relocating just a few streets, Lily feels that the move destroyed her.[92] Jim echoes the sentiment, and despite the lovely new houses, he concludes that for the older generation the demolitions 'nearly broke their heart'.[93]

Others also reveal how even a small move could be unpalatable. Ron's in-laws moved to a new house on the north side of Hessle Road when Subway Street, to the south, was pulled down. They hated the move and Ron explains why. 'They'd lived [*in Subway Street*] a long time,' he reflects, 'and my father-in-law was a fisherman and his father was a fisherman and her father was a fisherman and they were right in the hub of the fishing community. Subway Street, West Dock Avenue and all that there – that was just the hub of it.' When demolition took place, not everyone was relocated to the same area and so, says Ron, 'they lost friends and the days of standing on your doorstep and talking to your neighbour, as they used to do in the terraced houses … it just became a different way, a different community.'[94] Ron's family eventually settled, but others could not. In *Dear Tom*, the autobiography of Hull actor Tom Courtney, Tom's mother Annie expresses her restlessness in letters to her son. As the wife of a fish dock painter, she describes the move from the cramped, gossipy terraces of Harrow Street to the fresh air of Longhill with a growing sense of isolation. The distance to the fish dock meant that her husband spent more time travelling, and Annie missed familiar faces and communal contact. She expresses her disquietude as 'the most awful feeling of wanting to go home' and makes plans to return to Hessle Road.[95]

Thomas and Margaret both went to great lengths to remain in Trawlertown. Margaret was forced to move from several streets ahead of the bulldozer, before relocating to Boothferry Estate, away from Hessle Road. Margaret enjoyed her new bathroom and the garden for the children, but says, 'There wasn't the atmosphere.'[96] As a child Thomas and his family were moved from Hessle Road to one of the pre-war estates on Bricknell Avenue. Thomas hated it and as soon as he started fishing, he bought a house down Eton Street, back near the fish dock. These houses

92. Interview with Lily.
93. Interview with Jim.
94. Interview with Ron.
95. T. Courtney, *Dear Tom: Letters from Home* (London: Doubleday, 2000), p.241.
96. Interview with Margaret.

too were soon pulled down and Thomas had to move several times, each time returning to Hessle Road.[97] For those who did move away there were new terms of social engagement. Some narrators detected a change in the attitude amongst the neighbours who relocated with them to the new estates. Baths, gardens and fresh air seemed to bring with them a degree of separation.

Narratives of housing redevelopment on Hessle Road share commonalities with post-war developments in other working-class cities and neighbourhoods.[98] However, the presence of the fishery brought an additional dimension. Embedded within the locality were networks and affiliations rooted in the practice of an extreme occupation. Narrators do not always overtly connect their feelings about the disruption of Hessle Road to the fishing industry. Yet their narratives hint at an attachment that was particularly hard to break. The redevelopment of Hessle Road was a first step in the erosion of Trawlertown. The new homes brought progress, and with it more sanitary living conditions. But as Alan H reflects, 'it destroyed a community really – they distributed people into another place, which hadn't got the same deep-rooted sentiments'.[99] As a trawler skipper, Jim found the bond between the fishing community and Hessle Road to be unsurprising. Jim respected other dangerous occupations, like mining and steelwork. But fishing, he says, was different. 'At the end of their working day … they went home to their families,' he explains. 'We didn't. Some of us went and never came back. And the mothers, aunties, the children, they all helped each other. Well, you don't walk away and ever forget that kind of community.'[100]

Albert and William Wright Dock: An Untimely Beginning

As Hessle Road streets were redeveloped, life on St Andrew's Fish Dock continued as usual. The freezer fleet had expanded from the early 1960s, bringing with it new patterns of operation. Yet into the 1970s, the still sizeable sidewinder fleet maintained the old rhythms, and while some noted the appearance of cold stores for frozen fish, narrators did not generally attribute changes on Hessle Road at this time to the freezer vessels. It was

97. Interview with Thomas.
98. For example, Roberts, *Women and Families*, pp.211–221.
99. Interview with Alan H.
100. Interview with Jim.

perhaps too soon for their impact to be noticed, or the demolitions had masked other change. On St Andrew's Fish Dock, the freezer fleet initially required only small-scale adjustments. Frozen fish blocks could not be swung from trawlers in baskets in the wet fish tradition. Working as a fish dock bobber, John C. remembers the installation of a new system, which he calls a 'banana rig', that could land the fish mechanically.[101] A space was cleared to provide a concrete landing where two to three freezer trawlers could be unloaded and fish laid out before being transferred to the newly built cold stores.[102] In some cases, the cold stores had replaced the terraced houses to the south of Hessle Road; a literal example of the impact of the freezer vessels upon the surrounding district.

As the freezer fleet grew, more extensive adaptation was required. St Andrew's Dock was becoming outdated and its facilities dilapidated. From a management perspective, it was time to re-evaluate. Alan J. explains:

> The facilities for landing the ships weren't fantastic on St Andrew's Dock. We'd built a separate area for the freezer trawlers, but the wet fish trawlers still had lots of men working in the fish room and swinging the fish up using derricks ... The ships were getting bigger and there was a need to get better facilities ... a lot of the owners wanted to mechanise the landing of the ships.[103]

In June 1972, G. Maunsell and Partners were commissioned by the HFVOA to investigate the merits of relocating the trawling fleet to the adjacent Albert and William Wright Docks.[104] The Albert and William Wright Docks (henceforth Albert Dock) had been closed to commercial traffic in that year, despite having undergone a programme of modernisation less than ten years earlier.[105] These docks offered a more extensive water area, 3,453 metres of operational quay and an entrance lock that could admit the larger freezer ships. With a 60 per cent government grant, Albert Dock was converted to accommodate the fishing industry at a cost of £1.06 million.[106] In October 1975, the trawling fleet left their home of over ninety years and

101. Interview with John C.
102. Interview with Chris.
103. Interview with Alan J.
104. C/DBTH/15/129. Re-location of Hull Fishing Fleet.
105. SFIA, *Hull Fish Industry in 1983*, p.25.
106. SFIA, *Hull Fish Industry in 1983*, p.25.

Figure 13. The new fish market on Albert Dock in the early 1990s (Copyright Paul Hurst).

on 3 November 1975, Boyd Line's *Arctic Raider* was the last ship to exit, leaving St Andrew's Fish Dock redundant.[107]

The new fish dock on Albert Dock offered improved facilities, which in 1983 were hailed as amongst the most up to date and best maintained in the UK.[108] Despite initial disagreement between individual trawling firms and with the bobber's union, fish landing at Albert Dock abandoned the traditional swinging baskets in favour of a fully mechanised system of conveyor belts.[109] For trawling firms, mechanisation meant less labour, easier sorting and a better end product that met industry demands and EEC standards.[110] To bobber John C., however, it was a change for the worse. He felt mechanical landing was slower and under the new

107. Thompson, *A Tribute*, p.14. The SFIA Analysis of 1983 gives the date of the move as 1976. However, Thompson's date is supported by HCA (HC) Ref: C/DBTH/15/39. HFVOA. Minutes of a Meeting of the Management Committee on 24 September 1975. Item MC/97/75 and subsequent minutes.

108. SFIA, *Hull Fish Industry in 1983*, p.27.

109. For disputes over landing, see C/DBTH/15/39. HFVOA minutes from January 1975 to November 1976. For final outcome see SFIA, *Hull Fish Industry in 1983*, pp.29–30.

110. Interview with Alan J.

conditions he felt that he had to work longer hours for no extra pay.[111] For fish merchants, the move to Albert Dock was business as usual, but in a better working environment. 'Well, it wasn't a bad move in a way,' David explains, 'because the merchants moved into adapted warehousing, which [*had*] semi-individual stands. They were small units of three or four merchants and the actual fish was landed in separate warehouses. [*We were*] slightly more covered, as it were, from the elements.'[112] David and Chris agree that for the merchants the move brought improvements. But the modern facilities also diluted the concentration of activity that had characterised St Andrew's Fish Dock. In the summertime, David missed working outdoors and found that the warehouse environment felt enclosed, even unhealthy. It also altered the daily encounters that had formed the fishing taskscape. 'You didn't see your neighbours so much,' David recalls. 'I mean on St Andrew's Dock, when you were down on the quayside filleting, you could see people both sides for as far as the market went.' But on Albert Dock, high walls divided up the work areas or stands, and as David explains, it meant that 'you weren't aware so much of other people around about you'.[113]

Further separation stemmed from the fact that in the early years of the new fish dock, certain larger facilities, offices and other provisions continued to operate from the old St Andrew's estate. Dockside infrastructure that had developed over 90 years could not be relocated overnight. As Alan J. explains, when the fleet moved into Albert Dock, the ice company that supplied the wet fish trawlers remained on St Andrew's. 'We had to transport the ice on trucks,' Alan recalls, 'so it split [*things*] up, because some of the staff stayed in the old area. I mean the Fish Meal Company was right at the most westerly end of the [*old*] fish dock, y' know, you weren't going to move that.'[114] John C. continued to clock on at St Andrew's Dock and he and his fellow fish landers would be taken to Albert Dock by coach.[115] The dispersed operation and new landing practices altered the environment and experience of the dockside. Narrators observed the smaller size of the fish market, but also the erasure of the old dock's bustling character. 'It was in a couple o' warehouses,' Michael recalls, 'and the atmosphere

111. Interview with John C.
112. Interview with David.
113. Interview with David.
114. Interview with Alan J.
115. Interview with John C.

had gone completely.'[116] Thomas, who in later years worked in security at Albert Dock, missed the familiar spaces and collective activity of the past. 'There was no atmosphere there,' he explains, 'not like there was … y' see, everything happened on [*St Andrew's*] Fish Dock. Everything. Your trawlers were landed on Fish Dock, you signed on on Fish Dock, you signed off … So, to move it to another dock … it was all wrong.'[117]

Although commodities such as ice would eventually be produced on Albert Dock, events after 1976 ensured that it would never gain the status of St Andrew's as the heart of Trawlertown. The effort and expenditure on the relocation occurred with incredible mistiming, confirming just how unprepared the industry and government were for what was to follow. The planning and refurbishment of the new fish dock occurred in the midst of the turmoil surrounding the second Cod War. The transfer itself occurred on the very eve of the third. When the Albert Dock scheme was conceived there were, on average 4.25 trawlers landing each day (totalling c.597 tons of fish). By 1976 this had fallen to 2.4 (446 tons) and by 1983 it had plunged to 0.7 (c.85 tons).[118] The new fish dock was in trouble from the start.

Twice Disrupted

Given time, people and activity, Albert Dock would have generated its own taskscape, creating a new freezer-Trawlertown at the heart of a modern fishery. But as the industry moved to its new home, there came the final Cod War and the ensuing disruption. With ships laid up, sold or operating away from the city, the impact spilled from the dockside into a neighbourhood already unsettled by housing clearances. The residents of Trawlertown were hit by waves of transformation, first to their homes, streets and networks, then to the location and dynamic of their workplace. On top of that came the greatest threat of all: the contraction and loss of the fishery itself. At stake was the very existence of Trawlertown and the lifeworlds that it engendered.

From 1976, the impacts of the final Cod War and the emerging new ways of fishing were visible on the quayside. Fewer trawlers were arriving

116. Interview with Michael.
117. Interview with Thomas.
118. SFIA, *Hull Fish Industry in 1983*, p.25.

into Hull and when they did, it was with less frequency and from new waters. In line with the business model of the past, Hull trawler owners had taken on the full lease of Albert Dock and had invested in its facilities. By the early 1980s, they were battling to keep the dock viable. With the industry facing multiple challenges, the practice of running a substantial home base had become unprofitable. With only limited government assistance and wedged in a position of non-negotiation with the British Transport Docks Board, the trawler owners fought to keep the dock from closure. In February 1980, the HFVOA proceeded towards voluntary liquidation. Created in its place was the Humber Freezer Trawler Owners Company (HFTOC). Freed from earlier restrictions on rent negotiation, HFTOC strove for a better deal that could keep the dock operational. By the late 1970s, with fishing opportunities for Hull trawlers severely restricted, it became clear that if the dock and fish trades were to continue, Icelandic vessels must be attracted to land at the port. Occurring so soon after the forced withdrawal of British trawlers from Iceland, it was this move that had angered Thomas. Yet the landings were a necessity. Without Icelandic-caught fish, Hull's contracting fish trades could not have survived. The price was the growth once again of a risky dependency upon Iceland.[119]

As the Hull fleet landings declined and the industry struggled to source fish supplies, the effects could be seen in the vicinity of the dock. Engineering shops closed, along with the premises of joiners, riggers and plumbers.[120] The old icehouse closed in 1978,[121] but on the crumbling St Andrew's estate, some functions lingered on. Lily continued to work in her office by the lock gates and recalls that elements of the dock were still active when she left in 1981.[122] Amidst the growing dereliction, Ben remembers odd interim uses. 'There was a guy … where the old bank used to be,' he recalls. 'He was extracting gold off the old military equipment – the plate … He just set up for himself and he must've got the building for a song … y' know, he didn't need a penthouse.'[123]

119. Taken from various minutes, letters and supporting documents of HFVOA and HFTOC between July 1977 and April 1981. General papers, HCA (HC) Ref: DBTH/15/41; DBTH/15/42; DBTH/15/114; DBTH/15/121.

120. Interview with Philip.

121. Thompson, *Fish Dock*, p.70.

122. Interview with Lily.

123. Interview with Ben.

As the 1980s progressed and the industry continued to contract, the remnants of St Andrew's Dock hastened towards final closure. For many, it is the dereliction of this site, rather than the struggles of the newly established Albert Dock, that symbolise the end of the industry. Doug recalls the decay as he drove past the former Lord Line trawler company offices. 'It looked absolutely derelict,' he says, 'because the windows were smashed. Kids had obviously used it for a bit of entertainment on a night time.'[124] Lily describes her own return to St Andrew's Dock after 1981, saying, 'oh, it just looked derelict. It was awful … it was just so sad to see it … when I was always used to [*there*] being so many people about and so many things happening.'[125] The sight of a place once so animated, now reduced to silence and decay was an encounter that some chose to avoid. 'I went down there once … it just looked horrible,' Michael explains. 'No, as soon as fishing finished, I didn't want to know any more, cos it was depressing.'[126] Ken too confirms that he has spoken with former fishermen who refuse to visit the dock. 'They like to remember it exactly how it used to be,' he reports, 'and not see it in its dilapidated state now.'[127]

From the premises of the WFA, close by the lock pit, Ken had witnessed the transition from dockland to wasteland. With the dock gates closed and the strong daily tides, large holes had been cut in the lock gates to prevent silting. 'This didn't work at all,' says Ken, 'so people could see, slowly, the dock filling up and silting up with sand … and what nobody can fight against is Mother Nature, and grass and weeds and plants growing.' Ken watched through his office window:

> I could actually see it changing. You got people coming and throwing things in it. The strong tides used to bring all kinds of rubbish up the river and drop it off at the entrance to the lock, so it was a sorry state to see it. And then the trawling offices that were left … you would see all the windows getting broken and vandalised. And then one time, the demolition people came along, a big contract, to start to knock everything down.[128]

124. Interview with Doug.
125. Interview with Lily.
126. Interview with Michael.
127. Interview with Ken.
128. Interview with Ken.

From the mid-1980s, the site of St Andrew's Fish Dock began to be cleared and in-filled to accommodate new development (see Map 6).[129] Working now with the Humber pilots, Jim became part of the process. He describes the Dutch dredger that took sand from the River Humber, just outside Grimsby. 'It used to do that voyage twice a day,' Jim recalls. 'We had to board the pilots on and off each time and I saw that happening. It was a fine piece of engineering and work – no doubt about it – but it didn't soften the blow.'[130] Work was also underway on the construction of the A63, which was built on former dockland to the north of the dock basin and opened in November 1985. Former trawlermen returning to the site met with scenes of disruption. Around this time, Gill's husband John, with his newly acquired driving skills, decided to visit the old dock whilst on leave from the Gulf. The visit was something that John had done regularly, but on this occasion, he met with an unknown dirt track and a large digger. 'He drove down there,' Gill narrates, 'and I [*said*] to him, I think you've took the wrong turning … and this chap stopped him and he [*said*] – where you going Jack, you can't come along here … you'll have to go back.'[131] As the worker offered directions, John asserted, 'Oh I know my way around here.' But John was no longer on familiar ground. Demolition and rebuilding had swiftly rendered old haunts unrecognisable. For John it was disorienting and Gill recalls, 'he [*said*], I can't believe this. There's nothing here.'[132] By 1989, the majority of St Andrew's Fish Dock had been transformed into a retail zone. With just two trawling firms operating in the adjacent Albert Dock, the arrival of new shops and leisure units in place of ships and fish sheds confirmed the end. As Ken concludes, 'Fishermen would've seen cinemas going up and MacDonald's and all that and just think, well that's it'.[133]

129. Map 6 shows the retail estate of St Andrew's Quay, and to the east, the derelict buildings and basin of the St Andrew's Dock conservation area. To the north, the Clive Sullivan Way has replaced the lines of the railway. Beyond the carriageway, factory units and cold stores stand on the site of the former terraced housing.
130. Interview with Jim.
131. Interview with Gill.
132. Interview with Gill.
133. Interview with Ken.

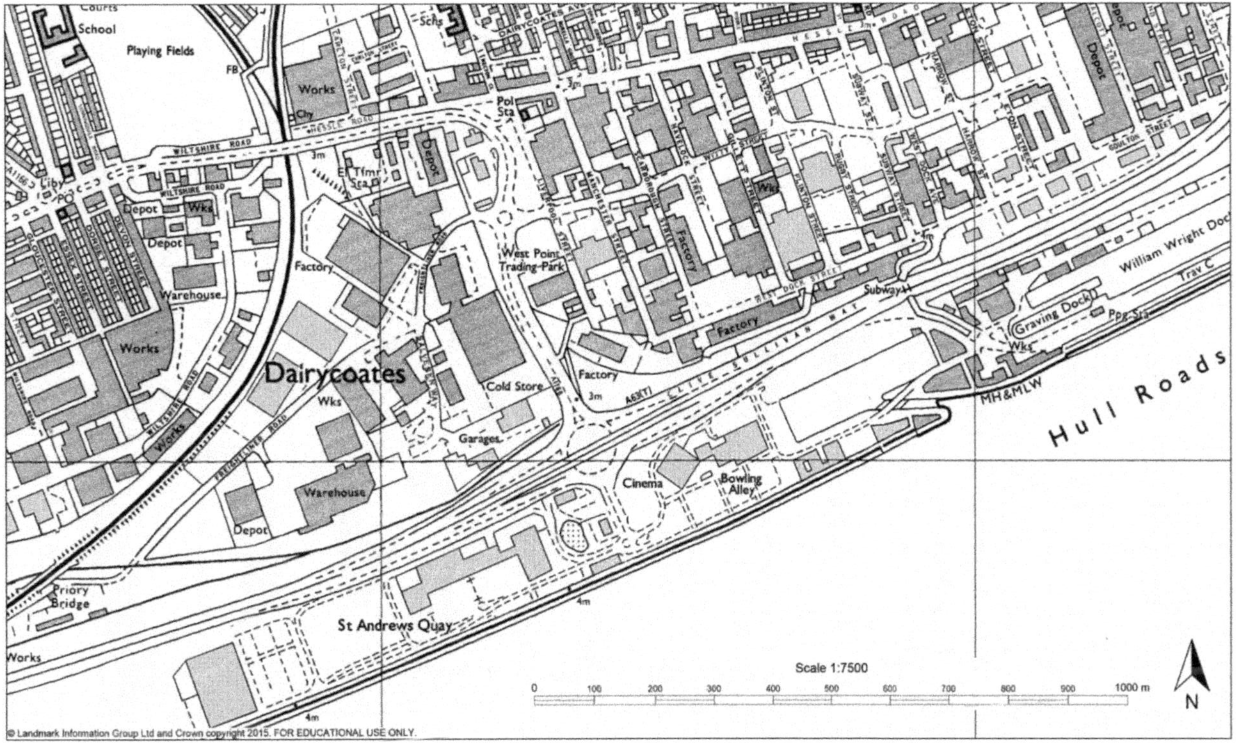

Map 6. 1994. St Andrew's Quay retail park and south side of Hessle Road.
Source: Landmark Information Group, *Kingston upon Hull, Yorkshire.* 1:10000. National Grid Tile TA02NE, EDINA Historic Digimap Service, 1994, http://digimap.edina.ac.uk (created 26 September 2015).

The Death of Trawlertown

Beyond the fish dock, Hessle Road too faced this new wave of disruption. With Trawlertown in its midst, the district not only supported fishing, it depended upon it. Less work within the fishery meant less money in pockets and less spending within local shops and pubs. Coming ashore as a deckhand, Ron would have put money into the local economy. 'For two days,' he says, 'it was enjoying yourself, doing things, spending money and keeping the businesses of the city going.' With the collapse of fishing, he observed, 'a lot of good businesses closed down'.[134] People too were departing. Hessle Road had always teemed with life, but the movement of inhabitants out to new estates and less people working in the depleted fishery altered the dynamic of the district. Doug, like many others, recalled that in earlier times, a visit to Hessle Road shops was a lengthy and sociable experience, with stops to chat with people that he knew. But in the 1970s, Doug found fewer familiar faces. '[*There were*] not so many people that you knew,' he reflects, 'not necessarily to talk to, [*but to*] look and say oh, that's so-and-so is that, just across the road there. And course that was because a lot of them had moved out to Bransholme and those sort o' places.'[135]

With the industry in rapid contraction and the catching sector in terminal decline, the fishing taskscape was dissolving. Trawlertown existed with the common activities, collective encounters, shared spaces and supportive networks of an extreme occupation. As Arctic trawling diminished and fishing became more global, ships and operations increasingly diverged from the city. Where they touched, they left a lighter footprint. The fishery that had dominated the landscape, economy and culture of Hessle Road for over a century would thenceforth influence a much smaller part of it. The departure of the sidewinders and reduction of the freezer fleet, the loss of ancillary trades and the contraction of fish processing, the decay of the old fish dock, along with housing clearances and the dispersal of inhabitants, all conspired to bring Trawlertown to an end. For as Ingold writes, 'the taskscape exists only so long as people are actually engaged in the activities of dwelling', and as Palmer observes, 'no sailortown could ultimately survive the decline of the port which had given it life'.[136]

In his narrative, Michael captures the demise of a trawling port city.

134. Interview with Ron.
135. Interview with Doug.
136. Palmer, 'Ports', p.148.

'I would say 1978 it was going down the tubes,' he recalls. 'There's all the fish houses gone and all the people who supplied kits and boxes to the market, … all that'd gone.' In the past, horses and carts and then fish lorries had driven up and down Hessle Road streets all day long. When this ceased, Michael says,

> It was like half of Hull had gone to sleep and just left … I suppose like California in the gold strike, when it'd gone and you just had towns with all the doors swinging empty and all that. And that's like Hessle Road was. All the factories closed down, cos they didn't need fitters for the trawlers anymore. That was it. It was horrible … And before, all the pubs used to be full at 11 o'clock in the morning and there was taxis darting all over the town. But all that'd gone see, all the atmosphere.[137]

This was the end of Trawlertown. As the fishery assumed a lower profile, the taskscape gradually dissolved into the undulations of the wider city. Much of Trawlertown had existed in the consciousness and connectedness of its inhabitants. Once these were gone it ceased to exist. In Trawlertown, the Star and Garter, known to locals as Rayner's, was widely recognised as the fishermen's pub. Beyond Trawlertown it is just a bar – the fisherman's pub only in memory.

However, within Ingold's concept, the taskscape can also find physical expression. 'A landscape', asserts Inglis, 'is the most solid appearance in which a history can declare itself.'[138] Ingold builds on this, reflecting, 'Thanks to their solidity, features in the landscape remain available for inspection long after the movement that gave rise to them has ceased.'[139] As the activities of the past give way to new encounters, physical residues are left behind. This is where the journey to heritage begins. The distinct interactions of Hull's fishery have gone and their expression in the landscape is eroded, but solid traces remain that are invested with the memories, meaning and attachments of those who were part of their creation and passing. In present landscapes they are valued or debated as sites of memory, as legacy and as tangible expressions of other lives.

137. Interview with Michael.

138. F. Inglis, 'Nation and Community: A Landscape and its Morality', *Sociological Review*, 25 (1977), p.489.

139. Ingold, 'Temporality of the Landscape', p.162.

Chapter Seven

Remembrance

The Pull of the Past: An Epilogue

> Still the old road, like a magnet, draws us back from near and far,
> Do the shopping Sat'day morning, have a pint in Rayner's bar.
> Lyrics from Keith Marsden's song 'Hessle Road'

There is a postscript to the demise of Trawlertown. As the song lyrics suggest, there has been a strong tradition for former residents to return to Hessle Road. Reference in a folk song confirms that the practice is recognised within the realms of popular memory, and among narrators there are those who report a sustained and conscious effort to retain a connection. In the era immediately after the housing clearances, some, like Ivy, would return to Hessle Road regularly to remedy a sense of homesickness. Others were drawn by familiar routines. Ivy's neighbours on the Bransholme estate, a former trawlerman and his wife, would visit Hessle Road weekly: she would have her hair done and do the shopping, whilst he would visit the pub and meet friends.[1] Over the decades, practices that were initially a response to the disruption of housing clearance and to the loss of the fishing industry have endured. Ken says that although many have passed on, there are still fishermen returning to Hessle Road 'to meet up and talk about old times'.[2] Ron reports, 'I just feel so comfortable when I go there.' He observes the new fabric and changing community but says 'that doesn't make any difference. I go down Hessle Road and Hessle Road is Hessle Road.'[3] Pauline, Victor

1. Interview with Ivy.
2. Interview with Ken.
3. Interview with Ron.

and Gill have all reported enjoying regular visits back. And Margaret, living once again in the area, asserts her absolute allegiance. 'I wouldn't leave here,' she says. 'If I won the lottery, I wouldn't leave Hessle Road.'[4]

Such returns can be seen as the continued pull of Trawlertown: the remnant rhythms of an extreme maritime occupation. Returning to Hessle Road is a performance of place attachment, a demonstration of affinity to a neighbourhood that shaped lives. The death of Trawlertown occurred with the divorce of a maritime community from the industry that was its creator. In existence, the intensity of its encounters built strong situated bonds. In demise, those same factors forged a desire for some to return and to remember. Remembrance is the final, yet ongoing narrative on this journey. Memory has so far provided a means to access a trawling history. In this chapter, memory becomes the subject; the focus of a sustained movement to acknowledge the past within the present-day city. As the collective practices of fishing dissolved, in their place arose a call to remember, represent and commemorate, as a fishery once worked and lived began its transition into heritage. The passage beyond Trawlertown is one of disrupted belonging, feelings of disinheritance, community dissonance and action against perceived acts of forgetting. Ultimately, it is a voyage of reclaimed identity, sprung from the never-ending messiness of remembering.

Understanding Memory and Place

> Haunted places are the only ones people can live in.
> Michel de Certeau, *The Practice of Everyday Life*

It is perhaps unsurprising that in this place-centred industry, the desire for remembrance has become firmly rooted in its former spaces. The merger of place and memory is fundamental to engaging with the past. Memory is inherently spatial, with the recollection of events, encounters and even feelings largely inseparable from the place in which they occurred. Places may therefore be valued because of their ability to connect with the past.[5] Where memory brushes against a landscape or townscape, it creates evocative terrain. When narrators recall Trawlertown, they are engaging with the geographies of memory. Like theories of place, geographies of memory

4. Interview with Margaret.
5. Scannell and Gifford, 'Psychological Benefits of Place Attachment', p.259.

fall between those that arise involuntarily from the practices of everyday living, and those perceived as social, political and cultural constructs. Involuntary memory is that which accumulates freely in the spaces around us. Smith presents memory, like heritage, as an intangible force, but it is one that is particularly powerful when it 'takes root' in the concrete.[6] Where memory gains physical expression the past can feel palpable. Casey's term 'place-memory' is memory that is 'place supported'.[7] Houses, streets and public buildings can all evoke personal memories or convey social or collective pasts. They hold an affective power, for as Creswell writes, 'it is one thing to read about the past in the book or see it displayed in a painting – it is quite another to enter the realm of history-in-place'.[8]

Some see the blending of memory and place as stemming from the practices, gestures and movements of ordinary life. Attention is now given to the significance of the mundane and everyday in our relationship with the past. Memory is seen as bursting beyond the museum and functioning in diverse, multiple and shifting terrains.[9] To de Certeau, the past lies beyond the tamed bricks and mortar of the restoration project. It is in the 'wordless histories of walking, dress, housing or cooking', and in the diverse expressions and narratives that 'turn the city into an immense memory where many poetics proliferate'.[10] The association of memory with practice and the everyday resonates with phenomenological and non-representational theories of place. Jones sees memory as integral to the process of becoming. People, he says, live in familiar environments and their existence within them is not merely performative, embodied and spatial, but 'temporal and memorial'.[11] Memory is essential to the ongoing creation of place, for 'we do not just "live in the moment" but in a progressing compendium of interacting lived moments' that we retain and carry forward.[12] To Jones, our spatial relationships of the present are therefore shaped by the past. Memory

6. Smith, *Uses of Heritage*, p.60.

7. T. Cresswell, *Place: A Short Introduction* (Oxford: Blackwell, 2004), p.85.

8. Cresswell, *Place*, p.86.

9. D. Atkinson, 'The Heritage of Mundane Places', in B. Graham and P. Howard (eds), *The Ashgate Research Companion to Heritage and Identity* (Aldershot: Ashgate, 2008), p.387.

10. M. de Certeau, *The Practice of Everyday Life. Vol. 2: Living and Cooking* (Minneapolis: University of Minnesota Press, 1998), pp.141–142.

11. O. Jones, 'Geography, Memory and Non-Representational Geographies', *Geography Compass*, 5:12 (2011), p.879.

12. Jones, 'Geography, Memory', p.880.

then, is 'a fundamental aspect of becoming, intimately entwined with space, affect, emotion, imagination and identity'.[13]

If it is through lived encounters that we build our connections with place, then through personal or collective memory, past affinities can be transported into the spaces of the present. De Certeau considers the landscapes of place-memory in terms of present absences; the shop that used to be a bakery, and the house where a friend once lived. 'What can be seen,' he reflects, 'designates what is no longer there.'[14] Yet just as the unstoppable flow of everyday life ensures that place is never fixed, the past is also not conducive to consistent remembering. Those who study the connections between memory and place find their subject unpredictable and responsive to the conditions of the present. With shifts of time, spaces of memory may be embraced then abandoned or reviled then cherished. They are constantly reinvested with meaning or they are simply ignored.

Beyond involuntary memory, there are other manifestations that derive from more structured calls to remember. Nora's classic concept of *les lieux de memoire* or 'sites of memory' is an engagement with the past that arises from the deliberate will to remember and is focused upon specific contexts; buildings, memorials, artefacts, even festivals and oral histories; anywhere memory is publicly invited to accumulate.[15] Memory here is an intentional act, a deliberate endeavour to anchor the past against what Nora refers to as 'the acceleration of history'.[16] The deliberate act of recall can push memory onto a political stage. Certain depictions of the past may be viewed as the manipulation of memory by elite or civic authorities, with the purpose of achieving power, stability, status or economic success. Just as with theories of place, such views conceptualise heritage as a socially constructed representation or text, embedded within a structure of power relationships. To some, statues, memorials, plaques, museums, buildings and the rituals that occur there create 'sanctified places of memory'.[17] The statements they convey may range from the ruthlessly political to the seemingly benign, but essentially, they privilege one figure, one community or one set of ideals

13. Jones, 'Geography, Memory', p.880.

14. M. de Certeau, *The Practice of Everyday Life* (California: University of California Press, 1984), p.108.

15. P. Nora, 'Between Memory and History: Les Lieux de Memoire', *Representations*, 26 (1989).

16. Nora, 'Between Memory and History', p.7.

17. K. Till, 'Places of Memory', in J. Agnew, K. Mitchell and G. Toal (eds), *A Companion to Political Geography* (Oxford: Blackwell, 2003), pp.289–290.

over the other. As a result, some memories endure a spatial absence, where the non-existence of physical reminders permits those in official positions to 'forget to remember'.[18] Commemoration is always partial, and in practice it holds the power to command attention or consign to oblivion. In the emotive landscapes of heritage, it is fertile ground for dispute.

Memory is further manipulated when pressed into the service of regeneration and development agendas. The heritage quarter, tourist event and place-promotion strategy make up Graham's 'external city', where selective extracts of the past are engaged to attract the attention of the outsider.[19] This commodification of heritage was derided in the 1980s as commercialised nostalgia or 'bogus history' that was rooted in conservatism and was counterproductive to progress.[20] Writers such as Raphael Samuel later reassessed aspects of commodification in more nuanced terms, placing them within a growing process of democratisation.[21] More recently, for example, Antonova and Rieser have demonstrated how the narratives or performances that are exchanged at touristic sites of heritage may both affirm and expand concepts of local and national identity, whilst also building economic and cultural resilience in the face of decline and transition.[22] Nevertheless, tourism and urban development have been identified as key triggers in the contestation of heritage.[23] The commercial use of the past can result in powerful memory dichotomies where perspectives, values and the need for profit may fail to align.

Official acts of remembrance also occur with the depiction of the past by the heritage curator. Smith's concept of the authorised heritage discourse, or AHD, conveys a professional discourse, which places the past in the hands of experts to be passively consumed by the public. Rooted in Western concepts of heritage, the AHD foregrounds the 'bounded' or the

18. M. Landzelius, 'Commemorative Dis(re)membering: Erasing Heritage, Spatializing Disinheritance', *Environment and Planning D: Society and Space*, 21 (2003), p.199.

19. B. Graham, 'Heritage as Knowledge: Capital or Culture?', *Urban Studies*, 39: 5–6 (2002), p.1009.

20. R. Hewison, *The Heritage Industry: Britain in a Climate of Decline* (London: Methuen, 1987); P. Wright, *On Living in an Old Country: The National Past in Contemporary Britain* (London: Verso, 1985).

21. Samuel, *Theatres of Memory*.

22. A. Antonova and A. Rieser, 'Curating Collapse: Performing Maritime Cultural Heritage in Iceland's Museums and Tours', *Maritime Studies*, 18 (2019).

23. Y. Yang, K. Dupre and X. Jin, 'A Systematic Review of Literature on Contested Heritage', *Current Issues in Tourism*, 24:4 (2021), p.458.

material – heritage as a site, an object or a building that can be preserved and managed. To Smith, the AHD naturalises and perpetuates the idea of inheritance, stewardship and expert care, which, she argues, serves to disengaging heritage from the active, cultural and emotional work of negotiating meanings from the past for individuals and communities.[24] The AHD may privilege singular narratives of nation (although it can occur at other scales), promoting the values and experience of the dominant social classes, whilst obscuring other forms of identity, and with or without intention, consigning some perspectives to silence.[25] With its focus on materiality and passive engagement, Smith asserts that it promotes a 'glass case' mentality – an inert and one-way process of instruction.[26] Yet whilst the AHD provides a theoretical language to examine the deployment of memory at formal sites of heritage, it does not reflect all curatorial approaches.[27] Smith herself acknowledges the growing body of heritage professionals and organisations seeking to engage with multiple pasts.[28] Movements to both assert and acknowledge the histories and heritage of diverse and local communities have made significant advances. With new attitudes, curatorial relationships have been revisited, narratives have been challenged and alternative expressions of memory have arisen 'from below'.

Post-industrial Acts of Forgetting

After 1976, the disruption in Hull's fishery occurred alongside transformations in other British manufacturing and maritime industries. Throughout Britain, de-industrialisation and changes in port practices disturbed once stable communities and rendered swathes of the urban landscape derelict. Within the maritime sector, Britain's relative performance was declining.[29] Containerisation and roll-on-roll-off transport had led to the development

24. Smith, *Uses of Heritage*, pp.29–34, Wetherell et al., 'Affective Heritage Practices', pp.7–8.

25. R. Mason and J. Sayner, 'Bringing Museal Silence into Focus: Eight Ways of Thinking about Silence in Museums', *International Journal of Heritage Studies*, 25:1 (2019).

26. Smith, *Uses of Heritage*, pp.29–34.

27. S. Wolferstan, 'An Ethnography of a 'Humble Expert': Experiencing Faro', in J. Schofield (ed.), *Who Needs Experts: Counter-Mapping Cultural Heritage* (Farnham: Ashgate, 2014), p.43; Muzaini and Minca, 'Rethinking Heritage', p.12.

28. Smith, *Uses of Heritage*, p.35.

29. Jamieson, 'Inevitable Decline?', pp.79–80.

of new container terminals that often repositioned the docks far from their long-established central city bases. The traditional port neighbourhood was approaching its end, leaving behind redundant and sometimes lucrative waterfronts in need of regeneration.[30] From the late 1960s, Hull's central 'town docks' faced closure and dereliction as the landing of commercial cargo relocated to facilities east of the city. In the west, Hull's fish dock enjoyed a more localised transfer to the neighbouring Albert Dock. However, the crises in trawling during the 1970s meant that ultimately the fishing district came to share the experience of other port city neighbourhoods. For as Palmer concludes, 'In this late twentieth-century world of maritime transport, the connection between city and port was to be finally severed, only to be rediscovered, or reinvented, as an aspect of urban heritage.'[31]

In post-industrial towns and cities, projects were sought to address the awkward abandoned spaces that embodied Britain's industrial past yet cried out for a new future. In the 1980s, the solution was often economy-driven. Conservation was combined with the active reuse of historic buildings, or learning merged with leisure in heritage themed tourist attractions. In such projects the past could become an asset. From heritage centres to dockside apartments, from family trails to place-marketing strategies, representations of the industrial and maritime past were harnessed to serve the economy. Yet such initiatives could face tensions, with community sense of place and experience at odds with the demands of a commercially driven heritage product. Where the past was pressed into the service of disparate commercial, civic and community needs, projects could struggle to meet differing expectations, resulting in the contestation that has led some to frame heritage as a dissonant discourse.

From the mid-1980s, as Hull's surviving fishery enjoyed a period of relative calm, Hull's trawling community faced a new dispute. Between 1985 and 1988, the derelict remains of much of St Andrew's Fish Dock had been demolished to make way for the first phase of a retail estate. Opening from 1990, the leisure and retail park offered a much-needed economic boost to the area. However, the redevelopment threatened to sweep away the remnants of Trawlertown. For many members of the beleaguered fishing community the demolition of the fish dock came to stand alongside other oversights as an act of cultural disinheritance. Scholars have highlighted

30. Palmer, 'Ports', p.150; L. Leeson, 'Our Land: Creative Approaches to the Redevelopment of London's Docklands', *International Journal of Heritage Studies*, 25:4 (2019), pp.365–367.
31. Palmer, 'Ports', p.150.

the apparent marginalisation of Hull's trawling legacy within the city's regeneration strategies in the 1980s and 1990s. Whilst other ports and cities were drawing upon their maritime and industrial heritage as a route to economic recovery, leaders in Hull brushed aside the trawling past. Lazenby and Starkey highlight a University of Hull student recruitment video issued in the 1980s entitled 'Everything but the Fish'. Here, Hull is promoted as a modern, vibrant city by overtly distinguishing this reality from the assumed grime and smell of a trawling port.[32] Trawling was also overlooked by strategic marketing consultants commissioned to draw investment to the city. Atkinson et al. have considered this omission in terms of urban boosterism, which dictates that to attract investment, marketing strategies must emphasise the positive elements of place, with adverse or commercially unattractive aspects of local identity, history or activity consigned to silence.[33]

Yet as Lazenby and Starkey observe, Hull's trawling history was not only distinctive and significant, it was 'replete with danger, excitement, distance and elemental force': attributes that they say 'make marketing people salivate'.[34] The attitude in Hull contrasts with the celebratory approach of rival trawling port Grimsby, where trawling heritage was embraced with the opening of the National Fishing Heritage Centre in 1991. To explain Hull's acts of forgetting, Lazenby and Starkey suggest a political desire, in the enterprise economy of the 1980s, to disassociate the city from an industry that from the 1960s had been portrayed as archaic and marked by poverty, social deprivation and low levels of educational attainment: a stigma enhanced by the 'decline and defeat' of the industry in the late 1970s.[35] With an urban and maritime economy more diverse than those of its Humber neighbour, to promote the city Hull leaders could simply look elsewhere.[36]

The physical fabric of St Andrew's Dock was also overlooked as a potential heritage asset. The ordinary nature of the dock made it easy to ignore. Attempts by the community to safeguard its buildings were hampered by a heritage protection system that held restricted and often nationally focused

32. Lazenby and Starkey, 'Altered Images', p.169.

33. D. Atkinson, S. Cooke and D. Spooner, 'Tales from the Riverbank: Place-Marketing and Maritime Heritages', *International Journal of Heritage Studies*, 8:1 (2002), pp.27–31.

34. Lazenby and Starkey, 'Altered Images', p.168.

35. Lazenby and Starkey, 'Altered Images', p.169.

36. Lazenby and Starkey, 'Altered Images', p.169.

definitions of significance and which placed emphasis on architectural merit. The prominent and successful dockside heritage developments of the time, such as St Katherine's Dock, London, and Albert Dock, Liverpool, had involved architecturally impressive structures. Opening in 1978, Hull's own award-winning historic warehouse regeneration project, the Waterfront hotel, had utilised a small but attractive brick building in a prominent city centre location. By contrast, the surviving fabric of St Andrew's Dock was domestic in scale, unexceptional in appearance and functional in nature. The modesty of the buildings obscured their significance as the home of Britain's once world-leading distant-water fishing fleet.

The potential for Hull's trawling narrative to thrive in the dynamic realms of heritage-led regeneration and place marketing was therefore not embraced. In a 1984 newspaper article, the leader of Hull City Council was asserting that 'fishing was never the dominating factor here'. Instead, he drew attention to the local presence of aerospace, engineering, pharmaceuticals and the new push for micro technology, concluding that 'we're … not forgetting the past, but we're also looking forward'.[37] By 1996, the civic elimination seemed complete, as the council leader once again reflected, 'while we are proud of our heritage, we must put the city's dominant fishy image behind us once and for all and move forward into a new era of pride and prosperity.'[38]

Yet amid such acts of forgetting, within Hull's fishing community there was a growing determination to remember. The desire for remembrance and representation can be seen as an afterlife to the intense activities and affiliations of Trawlertown. Phenomenological and more-than-representational concepts of place reveal a world that is produced through embodied practice and performance. Encounters and sensations from the past are part of this world, carried into the spatial relationships of the present via our always-at-work memory.[39] Around the streets of Hessle Road, moments once lived but now recalled have cemented the bonds of place. In turn such bonds have driven a desire to commemorate and represent. When describing the working fish dock, respondents draw readily upon their memories of sound, smell, feelings and movement. Vibrant descriptions capture not only the frenzy of daily practice, but also the rawness of affect. There is chaos and movement; slime, ice, bustle, noise. There are smells of rope and fish,

37. Lazenby and Starkey, 'Altered Images', p.169.
38. Atkinson et al., 'Tales from the Riverbank', p.90.
39. Jones, 'Geography, Memory', p.880.

and the warmth of arrival. Michael actively recreates the sounds of riveters working on long-since departed trawlers – 'bang, bang, bang, bang, bang'.[40] From this it is clear that in the realms of memory, the sensation of the real-time occurrence remains active and available.

Through such potent remembering the dock gained a new significance as the legacy of those who had experienced it at first hand. This re-assignment of meaning may explain why Michael and others avoided visiting the dock in decline. Narrators reveal emotional connection and feelings of loss at the demise of their trawling heartland. Lily laments that a place defined by people and movement should be reduced 'to nothing'. She conjures de Certeau's 'present absences' as she describes the derelict Lord Line building as 'a sad little reminder of something that used to be good'.[41] To Victor the derelict dock is 'symbolic of a whole life gone'.[42] Jim conveys loss in terms of a bodily affect, going on to make an overt connection between the empty dock and his sense of heritage:

Jo (Researcher): And did you ever go back and see St Andrew's Dock when it was empty?
Jim: Too many times [spoken sadly].
Jo: How did that make you feel?
Jim: Aw. Brok'en hearted! [sad laugh]. I know that sounds silly, but er [short pause] our history had gone. Our heritage'd gone. Absolutely. Oh aye![43]

As the encounters of the past came to an end, geographies of memory were left in their wake. Embodied activity had not only formed the dockland; it had imbued its spaces with meaning. When Ken was asked how he had been affected by the physical demise of St Andrew's Dock, he chose to focus upon a large redundant capstan that was located close to the lock gates. For Ken the capstan had acquired personal and collective significance. 'The reason I felt this,' he explains, 'is [*that*] every time we came into the port, [*there*] was the routine of a tug starting to tow us and I always had to put our head rope on this capstan to straighten the ship up. So, every ship that sailed out of Hull had some kind of contact with that

40. Interview with Michael.
41. Interview with Lily.
42. Interview with Victor and Pauline.
43. Interview with Jim.

capstan.'[44] As much of the dock was demolished, the survival of the capstan gained importance to Ken. 'It's still there today,' he reflects. 'It was the first piece of the dock and possibly it'll be the last.'[45] As the fishing community began to come to terms with their altered circumstances, there emerged a new affiliation to the old fish dock as a substantial site of memory.[46] As the fabric of the dock came under threat, this new allegiance would spur some into the arena to fight for representation and remembrance.

The Rise of 'Heritage from Below'

In 1975, the Hull Museum of Fisheries and Shipping, established in 1912, moved from a small brick pavilion in Pickering Park, west of Hessle Road, to its new city centre home. The Town Docks Museum, now the Hull Maritime Museum, took residence in the impressive former nineteenth-century offices of the Hull Dock Company, and work began to create a gallery telling the story of Hull's distant-water trawl fishery. In letters to the HFVOA, the museum sought advice and artefacts, outlining their intention to take the trawling narrative from the nineteenth century to the modern era of freezer trawlers and echo sounders.[47] The fishing gallery opened in 1976, claiming its place alongside other exhibits of the city's whaling, shipping and trading past.

The launch of the new fishing gallery, just like the move to Albert Dock, narrowly preceded the collapse of distant-water trawling. Here again, there is little evidence that the demise was anticipated. The museum's letters to the HFVOA do not indicate a collection policy based upon a sense of threat. Modern artefacts were requested merely to offer a 'comprehensive picture' from past to present.[48] The new gallery took a technical look at the history

44. Interview with Ken.
45. Interview with Ken.
46. Hilary Orange describes the transitory period in which people come to terms with the connotations of industrial decline as a metaphorical 'burial'. She suggests that with the lapse of time, immediate loss passes into memory and industrial ruins shrug of their subtext of failure to re-emerge as symbols of pride. H. Orange, 'Industrial Archaeology: Its Place Within the Academic Discipline, the Public Realm and the Heritage Industry', *Industrial Archaeology Review*, XXX:2 (2008).
47. HCA (HC) C/DBTH/15/39. HFVOA. Agenda Item 10 for Meeting of the Management Committee on 26 March 1975 and 24 September 1975.
48. C/DBTH/15/39. HFVOA. 26 March 1975.

of fishing, concentrating upon vessels, techniques and the development of the industry. As this authorised city narrative entered the 1970s, the tone was measured yet confident. The 21-day trip was conveyed with certainty as *this is what we do* and the Cod Wars were addressed as recent turbulence in the continuum of the fishery.

Over forty years later, visitors may have been unaware that the fishing gallery had changed little since 1976. Until its closure for refurbishment in January 2020, new additions were minimal. An audio-visual recording of ex-fishermen talking about their lives had helped to animate the technical exhibits. A graphic added in 2011 compressed into a single panel the struggles of the industry in the wake of the final Cod War. Certain panels carried a notice that their detail required revision. Yet despite the present tense used in later displays, the dated nature of the gallery would not have been immediately obvious. The opening of the fishing gallery just as the British fleet was forced from Iceland meant that without intention, the ending of the gallery's narrative coincided with what was to become of end of Hull's traditional fishery. Without modification, the account of *what we do* slipped silently and seamlessly into the conveyance of *what we did*.

The long survival of the 1976 fishing gallery offered a unique insight into the city's sanctioned and taken for granted view of its trawling past and 'present' on the eve of collapse. Had the industry remained active, it is likely that the museum would have continued as the single, obvious and accepted depository for Hull's trawling heritage. But the fishery plunged into crisis. Its spaces and practices faced termination, and new social and economic projects designed to reinvigorate the city failed to embrace trawling history. The response was cultural resistance that erupted from outside the authorised discourse. As 'heritage from below', it flowed beyond the museum walls, around the streets of Hessle Road and to the silted basin of the old fish dock, where it emerged as alternative expressions, grassroots commemoration and outright protest.

In the first decade after the final Cod War, as the fishing community dealt with the immediate blows, the past had not been a pressing concern. But from the mid-1980s, as the crisis settled, a desire to remember began to grow. Local historians Alec Gill and Michael Thompson produced a number of local trawling history books around this time. However, a perhaps unexpected contribution came from the theatre. In the early 1980s, local playwright Rupert Creed was seeking an identity for his newly formed Remould Theatre Company. Focusing upon the oral history method as a means to create compelling and authentic theatre, Rupert selected Hull's fishing industry for his first venture.[49] *The Northern Trawl* is what Rupert describes as 'the community's own view of itself'.[50] Drawn from oral history interviews, the play depicts the fishing industry through community stories and voices. First performed in 1985, Rupert sees the play as a milestone in acknowledging the city's trawling past. He explains:

> the moment we started recording people, we got a real sense that this was the story that they felt needed to be told. There was a passion about talking about the fishing. And there was also that feeling … that Hull had been forgotten a bit. So, this group of people felt that their contribution and their history had been side lined, had been unacknowledged and particularly because they had such a high death rate … they felt that this was wrong and they wanted it rectifying.[51]

Rupert's plays, along with the local history publications formed a body of work that responded to a need for the past to be recognised.[52] Reminiscent of Samuel's 'little platoons', they offered expressions of community memory that added to or countered the authorised discourse.

Alec and Rupert both declare their 'guest status' within the fishing community. Their role was that of mediator and their work a co-production between artist or historian and community. However, in 1993, from inside the community, George McGee felt motivated to launch his own heritage project. 'I'd walked down Hessle Road,' he explains. 'I knew then that Hessle Road'd vanished and everything would vanish with it. And that's when I got the idea of a mural – getting all the streets, pubs, clubs, characters, the trawlers.'[53] George is a lifelong Hessle Roader. If you cut him in two, he says Hessle Road would be written inside. After a spell as a fish dock barrow boy, George worked all his life in the building trade. Nevertheless, he firmly equates Hessle Road with fishing. 'Well the fishing

49. Interview with Rupert Creed, April 2013.
50. Interview with Rupert Creed.
51. Interview with Rupert Creed.
52. *The Northern Trawl* was followed in 1992, by *Vital Spark*, addressing the 1969 triple trawler disaster, directed by Jon Oram and Rupert Creed and performed by local people. In 1998, Rupert led a further oral history project with Hull University Lifelong Learning students, researching the women's 1969 campaign for trawler safety, leading to the book and the play *Turning the Tide*. R. Creed, *Turning the Tide: The 1968 Trawler Tragedy and the Wives' Campaign for Safety* (Hull: Back Door Press, 1998).
53. Interview with George McGee, September 2013.

industry was Hessle Road,' he says, 'it was the fishing industry that held Hessle Road together.'[54] This is reflected in the mural that he arranged to be painted on a large wall at the Edinburgh Street Community Centre. Taking years to fund and complete, *The Spirit of Hessle Road* depicts the names of familiar streets and shops, alongside Hessle Road scenes and well-known local characters.

At its heart are the trawlers and a roll of honour to those lost at sea. During the project, George took former fishermen to give talks in schools and brought classes to view the mural. Yet the mural was primarily aimed at an older generation. *The Spirit of Hessle Road* is an intimate portrait that could only be achieved with an 'insider's' knowledge and understanding. It was, says George 'for the old Hessle Roaders, because Hessle Road had been destroyed for them'.[55]

As the 1980s progressed, the spillage of memory took a new turn. The drive to repair the local economy had become focused upon the redundant spaces of St Andrew's Fish Dock. Initially the dock had attracted little public attention. But by 1988, the fish meal factory had been wound down and the remaining dockland estate was in the process of being demolished.[56] As Trawlertown faced oblivion, former fisherman and bobber John C. became increasingly convinced that its memory should somehow be anchored. As early as 1985, John began a one-man campaign to have the contributions of Hull's fishing community recognised within the planned rebirth of the old fish dock site. 'Being in the fishing industry practically all me working life,' he explains, 'losing two brothers and the amount of men what'd got lost from the port, I thought there should be something done about it.'[57] John's original aim was modest. The A63 approach road was under construction immediately north of the old dock basin. He felt that naming the road St Andrew's Way or Memorial Road would act as a fitting tribute, and he approached local councillors for support. However, shortly before the naming decision was reached, Hull rugby legend Clive Sullivan became terminally ill.[58] 'There was an East Riding Councillor saying that he'd been inundated with requests to call [*the A63*] the Clive Sullivan Way,'

54. Interview with George.
55. Interview with George.
56. *Fish Trader*, 15 October 1988.
57. Interview with John C.
58. Clive Sullivan played for both Hull FC and Hull Kingston Rovers. In 1972, he was the first black captain of any British national sporting team. He died in Hull in 1985.

Figure 14. *The Spirit of Hessle Road* mural at Edinburgh Street Community Centre, 2013 (Copyright Jo Byrne).

John recalls. 'So, I rang him and said that I'd suggested it be named as a tribute to the men who'd been lost in the fishing industry. And he said, oh yes, but Clive Sullivan's done a lot for Hull, y' know. So, I said, yeah, I said, the fishing industry did as well.'[59]

As the A63 became the Clive Sullivan Way, John expressed his objection in letters to the *Hull Daily Mail* and turned his attention to building a detailed model of the fish dock. A neighbour eventually put him in touch with Alec Gill, who in a separate protest had opposed naming the redeveloped dock Fisherman's Wharf, which he felt reflected San Francisco or London and was detached from local circumstances. Alec's own protests led to a local naming competition and the leisure and retail park became St Andrew's Quay.

The ensuing partnership between John and Alec led to the formation of the St Andrew's Fish Dock Heritage Park Action Group (STAND) at a public meeting held at Edinburgh Street School on Hessle Road in September 1989.[60] The group consisted of local people, many of them connected with trawling. It had three initial aims: to save the surviving fish

59. Interview with John C.
60. Interviews with Alec Gill, June 2013 and John C.

Figure 15. Looking east from the centre of the retail park, 2021. In the late 1980s, following protest and a local competition, the development was named St Andrew's Quay. The lone dockside building to the right is now a pub and is the only surviving structure outside the fish dock conservation area (Copyright Jo Byrne).

dock buildings; to construct a trawling heritage park; and as part of this to create a garden of remembrance dedicated to the lost trawlermen of Hull.[61] John adapted his model to show how a heritage park might look, and asserting their 'heritage from below', the close-knit ranks of Hull's fishing community were once again mobilised – this time to convince Grosvenor Square Properties Group plc (the property arm of Associated British Ports (ABP) and owners of the site) to preserve the buildings around the fish dock entrance. As the campaign grew, in contrast to the plays, books and murals, STAND's campaign became a direct challenge to civic decision makers and to recession economics. 'Heritage from below' had become heritage activism and it was driven by a sense of cultural disinheritance.

61. Alec Gill private papers. Document entitled 'The Purpose of STAND', undated.

Saving the Fish Dock

The demolition and redevelopment of St Andrew's Dock had commenced from the west, leaving the eastern area around the lock gates untouched. In December 1990, responding to STAND's campaign, Hull City Council had designated the cluster of buildings at the east end of the dock as a conservation area. Earlier that year, the Department of the Environment had rejected STAND's request to include certain dock buildings within the national statutory list, on the grounds that they were not of sufficient architectural or historic interest.[62] However, by that time, STAND had launched a petition that had attracted over 20,000 signatures.[63] In the days before e-petitioning, the figure demonstrated significant support for saving the fish dock. In local terms, it was hard to ignore.

The city council were not actively against the STAND proposals. By October 1989, several city councillors had expressed their support, followed by West Hull MP Stuart Randall and a number of Humberside county councillors.[64] But in the difficult economic climate of the 1980s, Hull City Council could not extend financial help.[65] The deputy leader, whilst reiterating the city council's general support, questioned the financial viability of the project, highlighting the 'complex considerations' that ran beyond merely seeking to save historic buildings.[66] One such complexity was ownership. The fish dock site was owned by ABP and local authority intervention was therefore restricted. Whilst the council were willing to facilitate, they were not willing to take on the project in its entirety. In a short newspaper article in November 1989, the city leader set out the position.[67] If STAND could persuade Grosvenor Properties to accept their plan, the authority would assist. Responding to criticism that fishing heritage had been ignored, the leader pointed to the authorised past, highlighting the contribution of the Town Docks Museum. The council could do no more, he concluded, they simply did not own the land.

62. Alec Gill private papers. Letter to Alec Gill from Department of the Environment dated 20 August 1990.

63. Interview with Alec.

64. Alec Gill private papers. Letter to Alec Gill from Councillor K. Branson of Hull Labour Group, dated 5 October, 1989; *Hull and Beverley Independent*, 17 November 1989.

65. Letter from K. Branson, 5 October 1989.

66. Alec Gill private papers. Letter to John Crimlis from Councillor P Bloom, Hull City Council, dated 31 October 1989.

67. *West Hull Target*, 9 November 1989.

The local authority's position put them at odds with the campaigners. However, the council's move to designate the fish dock as a conservation area had marked the site as a heritage asset. By 1992, Hull's £37.5 million bid to the government's City Challenge Scheme included a local authority and private and voluntary sector partnership that would convert the derelict spaces of St Andrew's Dock into a Marine Science Technology Park. The proposal incorporated a heritage museum, which would be located in refurbished dockside premises and run by the voluntary sector. The planned museum was an important indication that the dock's historical significance had now been accepted. When the Challenge Bid failed, the plans were transferred into a successful bid to the government's Single Regeneration Budget. In 1997, Hull City Council refused to grant planning permission to Grosvenor Waterside (formerly Grosvenor Properties) for a mixed leisure, retail and office development within the conservation area, on the grounds that it did not sufficiently address fishing heritage.[68] Although permission was later granted on appeal by a planning inspectorate unable to reconcile community value with nationally prescribed definitions of heritage merit, the council's initial decision was notable.[69] It confirmed the extent to which heritage prompted 'from below' had become embedded within the city's official discourse.

Despite protests, permissions, rejections and appeals, a revival of the fish dock as heritage was never to materialise. Over several decades, the site has faced disappointments and inconsistencies. New owners and unimplemented schemes have come and gone. In 2012, in a reversal of that earlier recognition of heritage value, the planning committee granted permission to Manor Property Group for a scheme that included the demolition of a substantial cluster of smaller structures within the conservation area.[70] With the buildings now gone, there is no sign of the redevelopment that sought their removal.[71] Their loss has pulled much of the heart and meaning from what is left of the dock. The remnant dock basin is barely discernible above the debris and silt that Ken first saw accumulating in the 1970s. On the quayside, the former Lord Line offices, Insurance Buildings and the Sea

68. Atkinson et al., 'Tales from the Riverbank', pp.31–32.
69. Atkinson et al., 'Tales from the Riverbank', pp.31–32.
70. Hull City Council, Planning Application Ref: 00025408N and 00025408M, received 23 October 2012. Permission granted at Planning Committee Meeting on 5 December 2012. Meeting attended by author.
71. Accurate at time of writing.

Figure 16. The Lord Line building, St Andrew's Dock conservation area, in an advanced state of decay, 2021 (Copyright Jo Byrne).

Figure 17. Derelict office buildings, St Andrew's Dock conservation area, 2021. Overlooking the River Humber, they are among the last survivors of a cluster of structures that once stood around the lock gates (Copyright Jo Byrne).

Figure 18. A sense of placelessness: looking west towards St Andrew's Quay from the conservation area, 2021. The crisp white sheds of the retail park stand in stark contrast to the silt and rubble of its neglected neighbour (Copyright Jo Byrne).

Fish Authority offices, along with the old pump house – the only listed structure on the site – are the survivors.

Exposed, isolated and in an advanced state of decay, they are more distant than ever from the bustle of Trawlertown. In development terms the site is challenging; a small tract of land containing redundant port infrastructure, bound by an estuary, a major road, a retail park and a working dock. It is hard to see a way forward and the answer may now be to reimagine, not rescue, the fragmented fabric of the past.[72]

Not everyone accepts that the dock is worth saving. The Lord Line building has been embraced by some as an emblem of Trawlertown; but for others it represents the trawling companies, not fishermen and their families.[73] In a letter to the local paper in 1998, one city resident complained that in his opinion the Lord Line building 'has damaged Hull's skyline long enough. Hull needs exciting new developments, not painful memories

72. C. DeSilvey, *Curated Decay: Heritage Beyond Saving* (Minneapolis: University of Minneapolis Press, 2017), pp.186–187.

73. Atkinson et al., 'Tales from the Riverbank', pp.36–38.

from a generally bitter, sad and hard life for thousands of Hull folk.'[74] Yet for many, the fish dock remains as their chosen site of memory, and there is discontent with the situation. For John C., the original campaigner, the outcome was a disappointment. On reflection, he concludes, it's sacrilege what they've done to that dock … and what they haven't done, y' know, what they should've done.' John would have liked to see the dock become a historical attraction, something akin to Beamish, the open-air museum of northern life, based in County Durham. 'Everything was there,' he says. But in John's view, 'They just let the buildings go to rack and ruin, so that they'd have to pull them down in the end.'[75]

A few narrators have suggested that the derelict site reflects the marginal position that the fishing community has held within the city, turning once again to the sense of 'otherness' that in the past was integral to life in Trawlertown. Ken and his friends have tried to keep the past alive in their own independent ways, giving talks and fish-gutting demonstrations at local events.[76] Such engagements take place away from the crumbling spaces of the old St Andrew's quayside. To many, the decaying dock remains a source of regret. As Michael concludes, 'For what the fishing industry did for Hull … and the money that [*trawlermen*] brought into the town. I think that the town could have made a better effort.'[77]

A Place to Remember

> But against the water's nice … Your memories are against water.
>
> Hessle Road resident reflecting on the site for a memorial

When speaking of fishing heritage, it is common for narrators to refer to the high death rate within the trawl industry. Loss is presented as their reason for remembrance, and the sentiment is echoed in the words of campaigners and in letters of support to the local press. The desire to preserve St Andrew's Dock sits seamlessly alongside calls for commemoration and a sense of duty to remember those that died. To Michael, the old

74. C. Wilson, in *Hull Daily Mail*, 20 January 1998. Atkinson et al., 'Tales from the Riverbank', pp.36–38.
75. Interview with John C.
76. Interview with John C.
77. Interview with Michael.

dock basin should have become a garden, containing 'a nice memorial to the fishermen who's gone, with all their names on, so their grandkids could see it'.[78] Feelings of loss continue to be a legacy of distant-water fishing. In the 140-year span of the Hull industry, 6,000 men are estimated to have lost their lives at sea.[79] Within the fishing community, it is recognised that for those who suffer this kind of bereavement, the absence of a body or grave is an additional and enduring source of grief. Stephens has revealed how 40 years after the collapse of trawling, the effects of losing a loved one at sea, coupled with the loss of the industry itself, still resonates within the pastoral work of the Fishermen's Mission in Hull.[80] That duality – loss of life, loss of industry – underpins the campaign for a tribute. One has folded into the other. The call for loss to be formally acknowledged gained pace once the industry itself was threatened with oblivion. On Hessle Road, families had supported each other and remembrance had been secured within the communal networks of the fishing community. As Trawlertown dissolved and its inhabitants dispersed, there arose more structured demands to remember. When John C. began his lobby, his purpose was to anchor the past amid a flood of change. Commemoration was part of that aim, and from conception STAND's plans for the fish dock had included a memorial.

It was Alec Gill who evolved the idea of an annual Lost Trawlerman's Day. His intention was to remember those lost at sea, whilst at the same time highlight the need for a permanent tribute. The first event took place in January 1990, timed to coincide with the most treacherous months of the trawling calendar.[81] It was located on the bullnose, a rounded area of land to the south of the decaying entrance to St Andrew's Dock. The site had become symbolic to the trawling community as the last point of contact with land for those who had sailed out of the city and had not returned.

In the early years, organising the event was a hands-on, community-led affair. Gill and other volunteers would help out, tidying up the dockside with spades and brushes. 'We used to have tea urns,' she recalls, 'and get

78. Interview with Michael.

79. This is the most widely circulated figure in media sources and political debate, as well as in popular discourse (although it can range from 5,000 to 8,000). It is unclear where the figure originated.

80. T. Stephens, 'Why Does Sudden Loss Continue to Feature So Much in the Current Pastoral Work of the Fishermen's Mission in Hull', Unpublished Undergraduate Dissertation, William Booth College and Havering College, 2012, pp.7–8.

81. Interview with Alec.

Figure 19. Royal Antediluvian Order of Buffaloes memorial located on the bullnose, 2009. In the background stands a cluster of derelict dockside buildings that have since been demolished (Copyright Hull City Council).

Figure 20. A place to remember: tributes to lost loved ones, 2021 (Copyright Jo Byrne).

the milk from Northern Dairies ... as a good will gesture. We used to get tea bags from the local grocery stores or coffee, y' know, everything was donated.'[82] From modest beginnings, Lost Trawlerman's Day grew to become an important annual occasion, attracting hundreds of participants, with support from the Hull City Council's event team.[83] By 1992, a small memorial plaque, donated by the Royal Antediluvian Order of Buffaloes, had been mounted on a simple plinth on the outer bullnose to provide a focal point.[84] However, STAND had hopes for a much larger and permanent tribute that would form part of the fish dock redevelopment. When the various schemes for regeneration failed, they set out to devise an alternative.

From 2003, STAND began to fundraise for their own independent memorial. Hull City Council assisted in land negotiations and contracts, but the plans were plagued by problems and delays. As the memorial process became protracted, alternative 'little platoons' of commemoration sprang up in its place. In Hull's Maritime Museum, the St Andrew's Memorial Books were commissioned by STAND, with the pages turned daily to reveal the names of those lost at sea on each date. A memorial corner in Hull Minster, created in association with the Fishermen's Mission, was similarly provided as a quiet space to remember as uncertainty hung over the dock. By 2010, the dilapidated state of the surviving fish dock buildings had forced Lost Trawlerman's Day away from its usual space. This was followed by the relocation of the planned memorial site itself. Ron, who was secretary of STAND until 2014, explains the circumstances. 'We wanted [*the memorial*] in the first instance on the bullnose,' he says. 'For quite a long period that was the allocated site.' However, with the area rapidly deteriorating and with no sign of rescue, they needed to explore new options. 'We came to the conclusion it had to remain on what was the St Andrew's Fish Dock,' Ron asserts, but they were forced to move away from the decaying bullnose and its health and safety challenges.[85]

A fresh location was selected on the River Humber frontage, further west and adjacent to the retail park, but within the area of the old dock. The move was emotionally charged. One group of ex-fishermen declared that

82. Interview with Gill.

83. St Andrew's Dock Heritage Park Action Group (STAND), https://www.hullfishingheritage.org.uk/lost-trawlermens-day (accessed 21 July 2020).

84. *Hull Daily Mail*, 27 January 1992.

85. Interview with Ron.

Figure 21. *St Celestine* leaving St Andrew's Fish Dock at the start of a 21-day trip in 1955 (Copyright Innes.co.uk).

Figure 22. The derelict lock gates and bullnose in 2021.
The site gained significance as the last point of contact with land for those who sailed from the city and did not return (Copyright Jo Byrne).

the proposed memorial would stand on the site of the old fish meal factory, and that it was 'an insult to anyone with relatives who were lost at sea to even consider remembering their loved ones there'.[86] But other objections reflected on the emotive power of the site that had been left behind.

Comments in the media returned to the bullnose as the preferred place to remember. One online petition read:

> The most fitting place for a Memorial is on Bullnose on St Andrews Dock which was often the last thing the men saw when sailing from Hull and the first thing they looked for on return to Port to know they were home back to their loved ones. Many have had their ashes scattered there and it is visited by many so they may feel close to the ones that have been lost or who have passed away.[87]

For some, the choice of memorial site was simply not negotiable. The change of location led to the emergence of an alternative campaign, driven again from within the community. The Hull Bullnose Heritage Group was formed in protest against STAND's forced surrender of the 'historic bullnose' as the designated memorial site.[88] Starting as a Facebook page, the group went on to develop their own plans for remembrance. At the heart of their campaign was the ongoing desire for a memorial located at the entrance to the dock. In 2015, when their initial plans for plaques and benches located on the bullnose once again met with difficulties, the group installed memorial benches and planters on Hessle Road. Other plaques have followed, in Rayner's, the former fishermen's pub, and in 2016 on the bullnose itself. In 2015, the group unveiled *The Last Trip*, their own freestanding memorial to lost trawlermen, made possible by local fundraising and located in a semi-permanent home in the central city. The Hull Bullnose Heritage Group retain the earnest hope that *The Last Trip*

86. *Hull Daily Mail*, www.hulldailymail.co.uk/Ex-fishermen-want-bullnose-rethink-St-Andrew-s-story-19609477-detail/story.html (accessed 12 February 2015).
87. Change.org, https://www.change.org/p/hull-city-council-we-the-hull-fishermens-history-pages-on-facebook-with-4430-members-call-upon-hull-city-council-to-reclaim-lord-line-building-bullnose-bostons-building-and-the-surrounding-area-back-to-their-control-and-provide-a-memorial-for-the-lost (accessed 11 February 2015).
88. Hull Bullnose Heritage Group, www.hullbullnoseheritagegroup.co.uk/?page_id=11 (accessed 28 July 2020).

Figure 23. For those in peril: the solid yet ethereal figures of STAND's memorial representing trawlermen lost at sea, 2017 (Copyright Jo Byrne).

will eventually transfer to their much-loved spot by the fish dock gates, overlooking the Humber Estuary. So far, it is yet to make that journey.[89]

In parallel, STAND's tireless campaign was coming to fruition. After a setback that saw an initial design for the memorial disputed within the community, in 2015, with a change of approach and a reduced cost, the group finally engaged local sculptor Peter Naylor to produce the successful concept.[90] Cast in bronze, Naylor's memorial depicts the silhouettes of 13 trawlerman, young and older, some with kit bags or waving in greeting or farewell. The figures are both solid and ethereal; present yet absent against the Humber skies. The piece was constructed by local firm W. Campbell & Son Ltd, and unveiled in a ceremony in February 2017. Framed by a

89. Accurate at time of writing.

90. STAND, https://www.hullfishingheritage.org.uk/memorial-heritage-garden/ (accessed 21 November 2020).

Trawlermen's Memorial Garden, it will provide a space at last for those who want to remember by the water.

There are now multiple tributes to the losses experienced over the lifetime of Hull's trawl industry. The tenacity shown by the fishing community in pursuing a memorial reflects the strength of the desire for remembrance. The dissonance surrounding its location testifies to the power of place. In Trawlertown, shared activity had formed bonds and built communal attachment to the dockside spaces. In relict form, those spaces threatened to divide, as place, memory and loss entered the contested domain of heritage and identity. Amid the emotional geographies of St Andrew's Dock, the place to remember became paramount and precise. Diverse memorials now collectively bring some closure in the long struggle to be acknowledged. For Peter Naylor, his sculpture is both a 'representational and figurative' means to connect people with this story. But ultimately, it is about the fishermen. 'The clue is in the name,' he says. 'This is a memorial to *the lost trawlermen of Hull*. Finally, we have brought them home.'[91]

Arctic Corsair*: Home from the Sea*

By 1986, as Hull's sidewinder fleet was scrapped, sold or converted to new use, the Boyd Line trawler *Arctic Corsair* became the sole survivor.[92] In 1978, the vessel had undergone an unusual conversion for pelagic trawling, and had been active in the south-west mackerel fishery.[93] The *Corsair* was laid up in 1981, but enjoyed a new lease of life after 1985, as British fishing entered a period of relative stability. In 1987, the trawler landed a record catch worth £170,000, before returning to port for the last time.[94] The idea of preserving a trawler had not been part of STAND's original plans. But by 1990, Adam Fowler had become the group's new chairman. With the *Corsair* laid up and rusting in Albert Dock, Adam recognised the opportunity and merit of saving Hull's last sidewinder. STAND launched a campaign asking Hull City Council to intervene, whilst approaching

91. P. Naylor, 'Design Description and Concept of the Pivotal Sculpture', https://www.hullfishingheritage.org.uk/memorial-heritage-garden/ (accessed 21 November 2020).

92. Thompson, *Hull's Side-Fishing Fleet*, p.94. The *Arctic Corsair* was built in 1960 in Beverley by Cook, Welton & Gemmell. By 1988, the trawler's registration had been closed and the vessel renamed *Arctic Cavalier*. The name *Arctic Corsair* was reinstated following purchase by Hull City Council in 1993.

93. *FN*, 31 March 1978.

94. *Hull Daily Mail*, 11 August 1993.

local business for support. With the impetus once more 'from below', the *Corsair* project proved to be viable. In 1993, the vessel was purchased by the council and restored under a partnership between the local authority, private business, training providers and volunteers.[95] The original intention was that the vessel would be the focal point of a St Andrew's Dock heritage centre. When the centre failed to materialise, the *Arctic Corsair* found a long-term home on the wharf of the River Hull, adjacent to the council's Museums Quarter.

Over the years, the partnership between the museum service, STAND and local volunteers has remained at the heart of the project. Local authority support has been crucial to success. The *Corsair* is owned by the council, who supply and manage budgets and secure grants for larger maintenance and presentation projects. Volunteers working for the museum give tours of the ship and carry out basic repairs. Until 2019, the council provided a stand-alone space within the Museums Quarter that functioned as a visitor unit. The content of the original unit was devised by STAND and later by museum service volunteers. In an act of democratisation brokered by the museum, the space was handed over to tell a local story.[96] The survival of the *Arctic Corsair* is a consequence of 'heritage from below'. In 1993, a *Hull Daily Mail* journalist captured the affective impact of the vessel. He writes, '[H]er devotees look on Hull's last sidewinder trawler with something like true love in their eyes. They can look past the dirt and the rust at what she once stood for.'[97] Affect stems from individual and collective conversations with the past. For those connected with fishing, the trawler enshrines their way of life. For others, it is an emblem of the city. Jim was formerly mate of the *Arctic Corsair* and became a member of the STAND campaign. He showed evident pride that the trawler was named alongside the Cutty Sark on the National Register of Historic Vessels.[98] Yet in the aftermath of trawling, the ship's significance was almost missed. Robin Diaper, Curator of Maritime and Social History at Hull Museums, suggests that the everyday presence of trawlers in the city had obscured their heritage value.[99] Hull's last sidewinder was rescued by those who understood its

95. *Hull Daily Mail*, 11 August 1993.
96. Interview with Ron.
97. *Hull Daily Mail*, 11 August 1993.
98. Williams, *Swinging the Lamp*, p.178.
99. Interview with Robin Diaper.

Figure 24. Back afloat: *Arctic Corsair* leaving the Museum Quarter berth on the River Hull, 2019, for a major refurbishment (Copyright Hull Maritime City, Hull City Council).

Figure 25. An impression of the *Arctic Corsair* in the purpose-built visitor centre at the North End Shipyard (Copyright Purcell Architecture Ltd).

meaning. Without their intervention, the *Arctic Corsair* would have faced the scrapyard – the unobserved victim of its own ordinariness.

In the autumn of 2019, the *Corsair* embarked upon a new journey. During that year, Hull City Council confirmed that it had secured a funding package of over £30 million to deliver the programme 'Hull: Yorkshire's Maritime City'. The programme draws attention to a diverse maritime past. Trawling is part of that narrative and in the regenerated dry dock and purpose-built visitor centre at the North End Shipyard, alongside the River Hull, the restored *Arctic Corsair* will have pride of place. With the trawler protected from damaging tides, visitors will walk on, in and under the ship, read about its past or listen to the volunteer guides, who remain integral to the experience.[100] It is unlikely that the *Arctic Corsair* will ever find its way to a heritage berth on St Andrew's Fish Dock. But in this new chapter of its story, the veteran trawler will find a long-term harbour at the very heart of the maritime city.

Dissonant Voices

The cultural legacies of Hull's distant-water trawl fishery owe much to 'heritage from below'. From the grassroots expressions of Rupert Creed's plays and George McGee's mural, to STAND's open protest and the dissonant campaigns of the Hull Bullnose Heritage Group, the former fishing community have been tenacious in their desire to remember. Through their determination, trawling has gained significant representation in the city's formal and informal heritage discourse. In 2017, when globally renowned artists were invited to respond to the city as part of Hull's term as UK City of Culture, stories of trawling were a prominent source of inspiration. However, the rise of fishing heritage is not universally celebrated. Some question what they see as a growing dominance of the fishing discourse, at the expense of other narratives. Dissent is evident within this Hull focus group – part of a project exploring local *spirit of place*:

Speaker A: To the outside world … that's how they see us. The fishing industry as the predominant industry.
Speaker B: It's nonsense.
Speaker A: It is nonsense. But … the fishing community really seriously over-romanticise the work that their forebears

100. Accurate at time of writing.

> did. And they do that very successfully. They've got a giant propaganda machine going there and no-one can get in to deny it and tell them the truth.[101]

Rick shares this outlook. As a musician, in the past he had experienced Hessle Road as a closed community, difficult to access for outsiders. He suggests that this internal cohesion remained at work in their campaigns for representation. It paints a partial picture. He explains:

> People say that I'm anti-fishing. I'm not. The problem I think is that fishing is related as being Hull. It isn't. Fishing is a part of Hull … I think that it's unfair that Hull is designated as being a fishing port. No, Hull is a port. What it did at one period of time, was in a certain area, it did fishing, and it's putting it into that context.[102]

In media comments online, those demanding that the city moves on have conflicted with those who call for remembrance. Ron responds to the critics:

> Yes, [*fishing*] has finished, and if this city's got to regenerate itself, it's got to look past that. But I still think that you should never forget it and it should be documented and it should be shown what an important part it played in the development of the city … I just think that when people say, oh forget it – we have some nasty comments sometimes – I just think people don't understand what it was about. If they say that it stood for nothing, then they don't understand what it was all about.[103]

Others have conveyed similar feelings that their past has not always been comprehended, and this perhaps is the crux of trawling's contested position within the city's heritage discourse. This localised industry and close-knit community – part of the city, but at the same time different – produced 'inside' and 'outside' perspectives. It built a sense of place that was lived, or merely observed. It bred the desire to remember, or people that 'don't understand'.

101. HCA Ref: C DSARC. *Hullness: The Search for the Spirit of Place*, Heritage Lottery Funded Project, Arc, 2011.
102. Interview with Rick.
103. Interview with Ron.

The Messiness of Remembering

As Hull's trawling networks collapsed, memory was left in the wake. Remembering is a present-centred act; something fluid, open to reinterpretation and responsive to the needs of the moment. Smith presents memory or heritage as a negotiation. It is, she says, 'a social and cultural process that mediates a sense of cultural, social and political change'.[104] Far from being in the past, this makes it 'vital and alive'.[105] Within Hull's fishery, the shift from active industry to maritime heritage has created a tapestry of memory, woven with lived encounters and place attachment, sewn with representation and memorialisation, and hemmed by acts of forgetting and resistance. Each has played a role in deciding the cultural legacy of a once cohesive community. In a context of maritime decline, these actions in synthesis bring insight into the processes of heritage and the complex 'messiness' of remembering.

The journey beyond Trawlertown connects everyday practice with the desire to remember. Narratives have illustrated how lived experience and collective, sensory and located encounters have built attachments to place that have in turn created emotive sites of memory. When these are threatened by demolition and loss, the same factors are evident in the desire to retain redundant dockland spaces as maritime heritage. The observation unites two typically divergent understandings of memory and of place. It shows that phenomenological and more-than-representational forces – practice, senses and feelings – are alive and influential as remembered acts in this community's campaign for cultural representation. In doing so, it reanimates singular theories that present heritage or place solely in terms of social or political 'texts' or representations, by reinstating the active human dimension.

The case study of Trawlertown also interrupts depictions of heritage dissonance that foreground national examples, and which emphasise conflicts between authorised, official or expert narratives and the communities that react against them. In the early days, STAND's heritage campaign *was* pitched against civic discourses and commercial strategies that they believed had sidelined community memory. Their campaign forced new approaches and alternative expressions that gradually secured a robust position for trawling heritage within the city. The achievement has been such that

104. Smith, *Uses of Heritage*, p.84.
105. Smith, *Uses of Heritage*, p.83.

some now question the dominance of fishing over other heritage narratives. In a cycle of cultural counter-hegemony,[106] the prominence of Hull's trawling past is now at times challenged by other cultural communities. The implication is significant and reinforces Robertson's idea that heritage is not only contested between those in authority and subordinate groups – dissonance is also evident *between peer groups*, where one is perceived to have gained cultural dominance.

Robertson's observation of the Lewis Islanders goes further. Here he shows that even at the most local level and from *within* a community, the merger of memory, memorialisation and place may be a potent source of dispute.[107] In Hull, just as on the Isle of Lewis, plans to remember trawling have met with disagreement from within the fishing community itself. Where this has been most powerful is in the context of commemoration. The tension between STAND and the Hull Bullnose Heritage Group over the siting of the memorial confirms not only the power of place, but also the intrinsic potential of heritage to be dissonant, even in a very local arena. Smith writes that 'heritage *is* dissonant'.[108] Disputes, protests and alternatives are central to the process of making meaning from the past. Dissonance and debate serve to confirm identities and keep the past relevant and alive. This does not only occur in national stories and between powerful elites and subjugated groups. It happens on our doorsteps, *between* and *within* our communities, in an ongoing and sometimes emotional dialogue with the local past.

At all levels, Robertson has conveyed the potential of 'heritage from below' to be disruptive.[109] However, in this story the relationship between 'heritage from below' and the authorised discourse calls into question the neat division between the two. In their campaign, STAND certainly challenged the civic authorities, asserting their right to define, protect and celebrate their own cultural heritage. Yet on closer inspection their activism was less subversive. The early focus was the civic naming of the A63 and St Andrews Quay, and on the designation of a conservation area that embedded the fish dock in a formal system of heritage legislation and protection. Campaigners later sought to transform the fish dock into an official heritage space, authorised to tell Hull's fishing story. In these

106. Robertson, 'Class, Social Protest', pp.146–147.
107. Robertson, 'Class, Social Protest', pp.146–147.
108. Smith, *Uses of Heritage*, p.82.
109. Robertson, *Heritage from Below*, p.23.

examples, there is an implication that whilst challenging the message of the city's heritage discourse, campaigners were seeking to be *part of it*. Narrators refer to Beamish or heritage projects in Hartlepool and Grimsby to highlight how the fish dock might have been. Jim expresses pride that the *Arctic Corsair* is included in the official National Register of Historic Vessels.[110] All this has implications for theories that rigidly divide heritage into expressions from above and from below. Muzaini and Minca highlight the often-overlooked potential for the authorised past and 'heritage from below' to entwine and overlap.[111] The result is a space of possibility. If STAND, whilst subverting the dominant discourse, was at the same time seeking a place within it, then it indicates a grassroots appetite for a sanctioned but *inclusive* order. Rather than radical disruption, it suggests that a position of influence *within*, rather than opposed to, authorised narratives may sometimes be an end goal. STAND's protest forced negotiation. In the case of the *Arctic Corsair*, it led to a successful partnership between museum and community. The outcome was a democratisation of the discourse – with the *Corsair* embedded within the museum's representation, but retaining its independence through the continued relationship with STAND and the museum volunteers. The hybrid is perhaps more sustainable in a local arena and it brings it owns dissonance and curatorial challenges. But it exceeds the polarities of heritage from above and from below, harnessing instead the promise of cooperation in creating a past where community voice is valued and memory can be mediated.

The Right Place: A Prologue

The negotiation of trawling memory now has a history of its own. In the 1970s, it was represented in the newly opened Town Docks Museum, and ignored in the spaces of 1980s economic regeneration. Memory has been contested – against city institutions and between and within communities. Alternative expressions have gathered momentum, and partnerships have been formed. The aim has been to find the 'right place' for Hull's trawling past, and in many ways this has been achieved. The journey has been long, accompanied at times by a sense of marginalisation and hurt. Yet fishing

110. Interview with Jim; see also J. Williams, *Swinging the Lamp: Nautical Memories of a Lifetime at Sea* (Hull: Riverhead, 2013), p.178.

111. Muzaini and Minca, 'Rethinking Heritage', pp.14, 17.

heritage has secured a strong presence within the city, with memorials, plays and artworks, as well as the *Arctic Corsair*, existing as visible testaments to one community's tenacious desire to remember.

Alongside tangible outcomes, there are other unacknowledged and perhaps unintended results. The lengthy campaign for a memorial, the tension over its location and debates in the local media, have served to sustain trawling memory, keeping it active and relevant in a manner that a fixed representation could not accomplish. At the same time, the memorial services and fundraising events organised by STAND, and later by the Hull Bullnose Heritage Group, have helped to maintain the identity of a fractured community. Such interventions are instances of intangible heritage. With their continuing activity, the campaigners have created a channel of memory and meaning running between members of a dispersed cultural community, between that community and the wider city and between one generation and the next. It is a notable achievement. Decades on from the Cod Wars, as the physical spaces of Trawlertown have crumbled, these surviving networks have succeeded in keeping the past alive.

In the meantime, the ruins of St Andrew's Fish Dock are trapped between past and future. Memory is entwined with the decay. As the struggles and disputes show, the dock is the 'right place' for remembrance, and some hold on to their hope that one day a fitting memorial will stand at its entrance. But a solution for the site remains elusive and the dock exists as an enduring sad reminder of 'something that used to be good'.[112] Away from the waterfront, the former galleries of Hull Maritime Museum have kept a quiet detachment from the memory struggles that took place outside their walls. In the twenty-first century, a new, refurbished museum will thread trawling stories through exhibits that explore sea and shore, work and home – exceeding the older narratives formed in the midst of the Cod Wars. Beyond the museum on Hessle Road, new murals and most recently, the community's own fishing heritage centre, set up in a local shop unit, persist in telling other versions of the past. The process has not reached an end. Remembering is the epilogue to this particular story of Trawlertown, but it is a prologue to conversations with the past that are still to come. In the ongoing negotiation, those conversations will decide what is recalled and what is relayed, in a city that lost, yet retained, its connection with fish.

112. Interview with Lily.

Chapter Eight

Beyond Trawlertown

The story of the British distant-water trawl fishery in the late twentieth century is one of disruption and transition. In the trawling port of Hull, the 1970s had opened with a long-established industry gradually adapting to new technologies, while holding on to the rhythms of the past. Despite the onset of neighbourhood and occupational change, the fishing community of Hessle Road had remained united by the centrality of trawling within their lives and culture. Commonality of experience, place, networks and perhaps outlook had built a close-knit world. Both Michael working as a skipper and Thomas as a skilled deckhand had started out like those before them as a galley boy or deckie learner. They shared an understanding of long hours, cold and fatigue, and had witnessed the dangers, but also the rewards, of life aboard a trawler. Ashore, they knew the challenges of fitting three weeks of living into three days. At home, Gill, Lily and Margaret, like other Hessle Road women, had watched husbands, fathers, sons and brothers depart and return. They had held their families together, sometimes working in the fish-processing factories, shops, bars or ancillary trades that underpinned the local economy.

By the end of the decade that coherence was difficult to discern. Fishermen now shared a new commonality of redundancy and an uncertain future. A glimpse beyond Trawlertown highlights the fragility of the interaction between human societies and the marine environment. It makes visible the local impact of global events. The story told here is one of a single British trawling port, but it is far from parochial. Hull's fishery developed and operated within international parameters. Place-centred business and cultural networks responding to the comings and goings of a home fleet had given the false impression that Hull's distant-water fishery was a local industry. In the turmoil of the 1970s, the international tenor of this 'local' business came into stark relief. Global policy and territorial waters brought change directly to Hull's doorstep. In the 1970s, the activity of the fleet was central to Hull as a trawling port, socially, economically and culturally. When global factors impelled new practices, the impact resonated through the city and the repercussions were felt in the working lives, physical spaces and neighbourhood networks of Hessle Road.

Contemplating Hull's post-Cod War fishery through oral history narratives brings a new dimension to economic, environmental and technical accounts. Cultural and community perspectives offer a different assertion of history. In business terms, the arrival of the freezer trawler brought efficiency but, with it, environmental threat. Community narratives, however, speak of new ways of being at sea and practices ashore that were slowly changing the experience of the port district. Cultural insights reveal that the new ships were not always welcomed by those raised on the 21-day trip. However, they were readily adopted by a younger generation, who were attracted by a safer and more structured trawling career, and who journeyed to Hull's fish dock from further afield than the home port.

Into an atmosphere of transition came turbulence. The decline of British distant-water trawling is often assigned simply to the Cod Wars. The reality is more complex. Throughout the 1970s, EEZs expanded rapidly and globally, shifting the parameters of distant-water fishing. In parallel, economic pressures, oil crises, overfishing and Britain's entry into the European Common Market converged into a perfect storm. Individually each might have been weathered, but together they spread turmoil. The focus on a single trawling port makes visible the collective impact of multiple challenges. Beyond fishing policy and fishing wars, local narratives reveal the struggle with uncertain guidance, shrinking seas and unpredictable quotas, firms and fishermen left to sink or swim, and ships scrapped or sold – all expressed with a sense that the industry was re-scaled with a human rather than fiscal cost.

A human perspective may also explain the response to the gathering storm. An entrepreneurial spirit had led to the rapid establishment of the Hull fishery in the nineteenth century and had sustained its interwar performance. Tunstall refers to the enthusiasm demonstrated by Hull trawler owners after 1945 in replenishing their fleets and adopting the new technologies of sonar and radar.[1] However, in line with other exponents of post-war maritime complacency, Tunstall condemns the owners' failure to embrace the early adoption of the freezer trawler, and directs harsh criticism towards conservative and fatalistic attitudes within the industry. Yet although business development and marketing models may have remained

1. Tunstall, *The Fishermen*. p.47.

moderate and responsive to the narrow demands of the British fish-eating public, in the 1960s and 1970s substantial technical innovation does filter into British distant-water trawling. Furthermore, following the crises of the mid-1970s, there is every evidence that Hull trawling firms reacted with speed, enterprise, ambition and adaptability in the face of the challenges. But clearly, the scale of the pending disruption had not been adequately foreseen. Here the explanation may be cultural as well as economic. In the past, steady business practices, traditional fishing grounds and domestic markets had served Hull trawling firms well. Innovation was risky, costly and intrinsically linked to supply and demand. Timing was key – a good idea at the wrong moment might yield a bad result. Until 'business as usual' could no longer be sustained, there seemed little reason to change. Fishing limits and a globalising industry forced rapid adjustment. Change occurred in many quarters, at the same time and with mixed results, creating a climate of adversity. With greater heed to the shifting global position on territorial seas that had escalated over decades, the industry might have anticipated and avoided the sense of crisis that hampered later efforts to adapt. But forward planning was not prevalent in trawling culture. In general, trawling companies retained short-term horizons, and many trawlermen, casually employed in a dangerous environment, were fatalistic. The trawling community was accustomed to living with fear and worry and the global situation could be easy to dismiss as just one more threat. Or perhaps it was not feasible to anticipate change on this scale. The need for controlled, territorial and sustainable fishing grounds is today unquestioned, but for many in the 1970s, the closure of open seas was unfathomable. As narrators reveal, it simply did not seem possible that nations could claim 200 miles of water as their own.

In the aftermath, Hull's once cohesive fishing community was forced to adapt to diverse new ways of living. For many trawlermen and trawl industry workers, this was a period of unemployment, underemployment and uncertainty. Recollections of men drinking on public benches and of whole families turning to benefits convey dysfunction and breakdown that for some stretched into the long term. Precarity is evident in futile journeys to Fleetwood in an attempt sign on a ship, in frequent job changes and in gambles taken across the globe. Yet these narratives also reveal resilience. Lazenby has called for a revision of the received British post-Cod War image of stagnation and social welfare.[2] The difficulty and disadvantage experienced

2. Lazenby, 'Change and Adaptation', p.195.

by British distant-water fishing communities must not be diminished, but they should also be set against other narratives. Contributors to this book, from all sectors of Hull's fishery, eventually found work, at sea, on shore or sometimes between the two. Maritime roles feature prominently, in home waters and overseas. Some were with new and remote employers, whilst others retained local ties, continuing to depend upon the initiative and efforts of local trawling firms. Ashore, the trawling fleet sustained a remnant secondary economy. At sea, Hull companies embarked on new ventures with their reliable Hull crews. Stories of surveying in India, mackerelling off the Cornish coast and trawling in New Zealand all involved Hull firms, ships and men. The difference was increasing distance. After 1976, with ships returning less frequently to the home port, those still serving at sea could no longer imagine that they were working in the city.

A port-based study reveals the additional divergence between the fish trades and catching sector. In the early 1970s, Hull's fishing fleet and the fish trades had thrived in synergy. Freezer trawlers, however, had enabled frozen fish blocks to be traded at long range as a commodity. As Hull's catching sector ran into difficulty, flexibility of supply became crucial. Imported frozen fish and wet fish brought from overland and overseas ensured that Hull could survive as a processing centre. Mechanised landing and cold stores – new technology developed for Hull trawlers – helped to secure that future. But it was no longer Hull ships that dominated supply. The shift brought vulnerability to international factors and a new dependency upon Iceland. But in the critical years after 1976, connecting international supply with local demand allowed Hull's fish trades to survive. For the catching sector, things were different. Despite substantial efforts, diversification could not sustain the majority of Hull's trawling firms and after 1985, only two fishing companies remained operational. A port-based study again suggests a reason. After 1976, new projects were trialled concurrently; mackerelling took place alongside experimental fishing in deep waters; forays into oil next to ventures in the South Pacific; fishing in North Africa amidst the strain of keeping Hull's new fish dock active. In stable times, losses from one new activity might have been absorbed. In the crisis environment of the late 1970s, the challenge of multiple struggling projects proved to be overwhelming.

After 1985, Hull's trawl industry enjoyed a modest recovery; its fish trades in particular gaining strength. Beyond the Cod Wars a question is, was the experience of Hull's fishery one of transition or decline, contraction or demise? The answer is that it was all these things. The slow transition towards a freezer fishery was overtaken by the rapid decline of the trawling fleet. However, the contraction of the fish trades was arrested by transition to new suppliers. Some fishing firms survived by successfully adapting to flexible fishing and new maritime operations, while others closed. Ashore on Hessle Road, change brought by housing policy merged with decline brought by the international politics of fishing. For those rendered long-term unemployed, the late 1970s brought an end, but for others who found fulfilling new careers, it was a transition. After the Cod Wars, the lives of all those connected with the industry were transformed. The close-knit world of the fish dock would never return. But although Hull's fishing community had faced loss, some things had been gained, and through the persistence of collective memory, the relict spaces of Trawlertown could still evoke a neighbourhood whose fortunes were tied to fish.

Alongside fishing, notions of place have been central to this story. The journey beyond Trawlertown has drawn upon phenomenological geography and more-than-representational insights to reveal the port city as a network of meaningful encounters. Visualising the human and spatial interactions of Hull's fishery as a taskscape, or Trawlertown, adds to a body of work that explores the port city in cultural terms. As a model it can transfer to other port districts. Trawlertown is companion to Hugill's generic Sailortown, Daunton's Tiger Bay and Portsmouth's 'sailorhoods'. Yet while sharing commonalities, conversely it is different, for Trawlertown was shaped by a lesser-known engagement with the sea and was the distinct product of the Arctic trawl.

In the late twentieth century, Trawlertown was transformed. Change was driven by local planning and global politics, but it was experienced directly on the streets of Hessle Road. The disruption of the fishery forced new interactions. In 1957, Horobin had warned of the social and economic dangers of disrupting the 'special circumstances' of this particular port city district.[3] Yet from 1976, the fishing community faced the triple impact of housing redevelopment, economic crisis and changing practices. The longstanding bond between industry, workforce and district was broken, and the defining rhythms and culture of Trawlertown faded into the wider city.

Today, a landscape of emotional space is sustained by local memory. How trawling has been remembered and represented is integral to its history. Understanding the industry's intense practices, close-knit networks and experience of rapid decline makes it easier to see why fishing heritage

3. Horobin, 'Community and Occupation', p.353.

remains so emotive and important. The campaigns of STAND and the Bullnose Heritage Group, along with the grassroots projects, partnerships, socials and ceremonies, all express affinity to the recent past and to the sites of memory that it has produced. Finding the right place for trawling heritage has exposed the messiness and dissonance of remembering. The story of Hull's fishing past has journeyed from the curated displays of the 1970s museum, through the contested spaces of 1980s urban regeneration. In the 1990s, it was championed 'from below' and in the twenty-first century, a more diverse trawling narrative will find its place within the reinterpreted displays of Hull Maritime Museum. The *Arctic Corsair*, with a dedicated team of volunteers, will stand at the heart of the maritime city as an enduring emblem of community collaboration and memory. And although the abandoned fish dock remains locked into dereliction, the decay itself commands attention. Beyond physical preservation, DeSilvey has highlighted alternative systems of significance, interpretation and memory making that may arise from gradual decay.[4] Uncertainty and debate over the untended dockside have sustained the conversation with the past, and as resurgent new artworks, plays and murals confirm, trawling memory remains alive and relevant within today's unfolding city.

If remembering is something that belongs to the present, then recent reflections offer a momentary end. Four decades beyond Trawlertown, some like Philip have pondered how life might have been without political or territorial seas.[5] With smaller ships or an agreement with Iceland, could there have been a future for distant-water trawling ports like Hull? John C., however, has contemplated the winter losses and the industry's high incidence of accidents and death, and wondered if decline could have been a blessing in disguise.[6] Yet despite the danger, loss and hardship, there are those who missed the distinct, familiar rhythms of a close-knit fishery. It is Gill who captures the dichotomy of living with and remembering trawling. She is unflinching and succinct in her obituary to an extreme maritime occupation. 'The sea was good to me,' she reflects, 'but it was bad to me as well.' 'It's my life,' she concludes. 'I'll never forget it.'[7]

4. DeSilvey, *Curated Decay*, pp.14–17.
5. Interview with Philip.
6. Interview with John C.
7. Interview with Gill.

Appendix 1

Table 1: The rise and fall of the Hull stern, freezer and factory fleet, 1961–86.

Year	Arrivals		Departures
	New ships	Ships transferred to Hull	
1961	1		
1962	1		
1963			
1964	2		
1965	3		
1966	7		1
1967	3		
1968	5		2
1969	3		
1970	1		
1971		1	
1972	4		
1973	7	7	6
1974	1		1
1975	3		1
1976			2
1977		2	6
1978		6	3
1979			2
1980			3
1981			3

Year	Arrivals		Departures
	New ships	Ships transferred to Hull	
1982			12
1983			2
1984			6
1985			7
1986			1

Note: Up to 1973 the fleet is expanding. The activity between 1973 and 1975 reflects a number of operational factors, including Ross Group's relocation of their freezer fleet to Grimsby. *Junella 1* was sold in 1973 and *Junella 2* entered the fleet in 1975. In 1962 *Junella 1* had been Britain's first whole fish freezer trawler. In 1975, as the industry went into decline, *Junella 2* was the last freezer trawler to be built for over a decade. From 1975 the figures show a steady decline.
Source: Thompson, *Stern Trawling Fleet*, pp.125–129.

Table 2: New roles for Hull stern, freezer and factory ships after 1975.

New roles[a]	Number
Standby	12
Non-European fisheries	11
Nordic fisheries	7
Survey (Marr)	4
Survey (other)	3
Other European fisheries	3
Diving support	3
Scrap	2
Other	1

[a] Includes ships sold or redeployed by Hull owners to work in other fisheries or maritime sectors.
Source: Thompson, *Stern Trawling Fleet*, p.84.

Table 3: Number of sidewinder trawlers registered to Hull, 1961–86.

Year	Number of Hull-registered sidewinder trawlers
January 1961	140
January 1966	115
January 1971	73
January 1976	36
January 1981	5
January 1986	1

Source: Thompson, *Hull's Side-Fishing Fleet*, p.94.

Appendix 2

Table 1: Employment figures in the Hull fishing industry, 1960–93.

	1960	1972/73	Mid-1976	1983	1993
Catching sector	3,000	2,307	1,850	480	336
Ancillary trades (non-fish)	2,850	Unavailable	2,900	1,193	781
Fish trades (incl. fish meal and oil)	3,100	Unavailable	3,850	2,822	2,474
Combined onshore trades	5,950	8,700	6,750	4,015	3,255
Total	8,950	11,000 (approx.)	8,600	4,495	3,591

Note: Precise and consistent employment data for the Hull trawl fishery are difficult to obtain. The industry employed many workers on a casual basis and the numbers engaged can exceed the listed of positions. The figures here are compiled from a range of sources, which have drawn upon differing criteria with variable accuracy. Although this makes precise chronological comparison impossible, the figures do reveal general employment trends over the study period.

Sources:
For 1960–83: Sea Fish Industry Authority, Industrial Development Unit (SFIA) and University of Hull, Department of Geography, *An Analysis of the Hull Fish Industry in 1983*. Technical Report No. 220, August 1983, pp.81–84; for 1993: D. A. Palfreman, *The Fishing Industry of Yorkshire and Humberside: A Regional Study*, commissioned by the Yorkshire and Humberside Consortium, 1993, p.12.
The figures from *Hull Fish Industry in 1983* are taken from the following sources:
1960: Ministry of Agriculture and Fisheries, reproduced from Tunstall, *The Fishermen*, p.271. The authors of the report class these figures as underestimated, as they exclude certain categories of ancillary workers, p.80.

1972–73: Various Department of Employment, local authority and employment census records. The report authors define the total as an approximation.
Mid-1976: Yorkshire and Humberside Economic Planning Board, 1976.
1983: Various, including interviews with leading firms, industry bodies and a postal questionnaire. The authors class the calculations as estimates and refer to the catching sector figure as an 'extremely variable total'.

Bibliography of Referenced Sources

Primary Sources

Oral History

Oral history interviews conducted by author between November 2012 and February 2015.

Newspapers and Journals

Fish Trader, 15 October 1988.

Fishing News (*FN*), January 1975 to August 1982.

Fishing News International (*FNI*), August 1976 to January 1977.

Hull and Beverley Independent, 17 November 1989.

Hull Daily Mail, 27 January 1992.

Hull Daily Mail, 11 August 1993.

Hull Daily Mail, 20 January 1998.

West Hull Target, 9 November 1989.

World Fishing, September 1977 to July 1989.

Primary Documents

Hull City Archives (Hamling Collection): Archive Ref: C/DBTH

British Fishing Federation: Archive Ref: C/DBTH/20

C/DBTH/20/69. Item 76/38 in Minutes of a Meeting of the Board of Directors on 4 March 1976.

C/DBTH/20/72. Item 77/55 in Minutes of a Meeting of the Board of Directors held on 3 March 1977.

C/DBTH/20/73. Item 77/181 Minutes of a Meeting of the Board of Directors on 6 October 1977.

C/DBTH/20/73. Memorandum, Agenda Item 6 for Meeting of the Board of Directors held on 1 December 1977.

C/DBTH/20/74. Item 78/02 in Minutes of a Meeting of the Board of Directors on 12 January 1978.

C/DBTH/20/74. Minutes of a Meeting of the Board of Directors held on 4 May 1978.

C/DBTH/20/75. Memorandum. Agenda Item 8 for Meeting of the Board of Directors on 5 October 1978.

C/DBTH/20/78. Agenda Item for Meeting of the Board of Directors held on 7 February 1980. Programme for Improving the Processing and Marketing of Fish and Fish Products in England and Wales under EEC Regulations 355/77.

C/DBTH/20/82. Item 82/66 Minutes of a Meeting of the Board of Directors on 29 September 1982.

Hull Fishing Vessel Owners Association: Archive Ref: C/DBTH/15

C/DBTH/15/123. HFVOA Cadetship Scheme c.1970. Document Project 70s, 21 September 1970.

C/DBTH/15/129. Preliminary Feasibility Study of Proposed Re-location of Hull Fishing Fleet. G. Maunsell and Partners, June 1972.

C/DBTH/15/39. Agenda Item 10 for Meeting of the Management Committee on 26 March 1975.

C/DBTH/15/39. Item MC/29/75(a)(i) in Minutes of a Meeting of the Management Committee on 26 March 1975.

C/DBTH/15/39. Agenda Item 2(a) for Meeting of the Management Committee on 23 April 1975.

C/DBTH/15/39. Agenda Item 10 for Meeting of the Management Committee on 24 September 1975.

C/DBTH/15/39. Minutes of a Meeting of the Management Committee on 24 September 1975. Item MC/97/75.

Humber Freezer Trawler Owners Company: Archive Ref: C/DBTH/15

C/DBTH/15/115. Minutes of a Meeting of Trawler Managers and Shipping Masters held on Thursday 19 November 1981.

C/DBTH/15/115. Minutes of a Meeting of Shipping Masters and Trawler Managers held on Thursday 26 November 1981.

C/DBTH/15/115. Letter from Registrar to all Directors dated 12 February 1982.

Other Documents

C/DBTH/59/21. The UK Fishing Fleet in the Mid-Seventies. White Fish Authority, Fishery Economic Research Unit, August 1969.

C/DBTH/70/28. Mackerel Research in the South-West. Ministry of Agriculture, Fisheries and Food: Directorate of Fisheries Research. Laboratory Leaflet No. 32. Lowestoft, 1976.

C/DBTH/70/32. Blue Whiting. Ministry of Agriculture, Fisheries and Food: Directorate of Fisheries Research. Laboratory Leaflet No. 45. Lowestoft, 1979.

C/DBTH/6/4/11. Various Papers Relating to Hamling's Interest in West Africa, including Mauritania, 1979–80.

Hull City Archives: Other

C DSARC. Papers relating to the Hullness Project: *Hullness: The Search for the Spirit of Place*, Heritage Lottery Funded Project, Arc, 2010–12.

HL711HE. City and County of Kingston upon Hull, Redevelopment of Decayed Area; with particular reference to Residential Area 17, H. E. Alton, May 1960.

Industry Reports

Palfreman, D. A., *The Fishing Industry of Yorkshire and Humberside: A Regional Study*. Commissioned by the Yorkshire and Humberside Consortium, 1993.

Sea Fish Industry Authority (SFIA), Industrial Development Unit, *The Importance of Klondyking to the UK Pelagic Fishery*. Internal Report No. 1272, May 1986.

Sea Fish Industry Authority (SFIA), Industrial Development Unit and University of Hull, Department of Geography, *An Analysis of the Hull Fish Industry in 1983*. Technical Report No. 220, August 1983.

Alec Gill Private Papers

Letter to Alec Gill from Councillor K. Branson, Hull Labour Group, dated 5 October 1989.

Letter to John Crimlis from Councillor P. Bloom, Hull City Council, dated 31 October 1989.

Letter to Alec Gill from Department of the Environment dated 20 August 1990.

'The Purpose of STAND', document, undated.

Secondary Sources

Books

Abrams, L., *Oral History Theory* (London: Routledge, 2010).

Courtney, T., *Dear Tom: Letters from Home* (London: Doubleday, 2000).

Creed, C., *Turning the Tide: The 1968 Trawler Tragedy and the Wives' Campaign for Safety* (Hull: Back Door Press, 1998).

Creed, T. and Hawkins, J., *The Northern Trawl* (Beverley: Hutton Press, 1986).

Cresswell, J., *Time and Tide: The Story of the Craig Group* (Aberdeen: The Craig Group, 2003).

Cresswell, T., *Place: A Short Introduction* (Oxford: Blackwell, 2004).

De Certeau, M., *The Practice of Everyday Life* (California: University of California Press, 1984).

De Certeau, M., *The Practice of Everyday Life. Vol. 2: Living and Cooking* (Minneapolis: University of Minnesota Press, 1998).

Dennis, N., Fernando, L., Henriques, M. and Slaughter, C., *Coal is our Life: An Analysis of a Yorkshire Mining Community* (London: Eyre & Spottiswoode, 1956).
DeSilvey, C., *Curated Decay: Heritage Beyond Saving* (Minneapolis: University of Minneapolis Press, 2017).
Dicks, B., *Heritage, Place and Community* (Cardiff: University of Wales Press, 2000).
Gilchrist, A., *Cod Wars and How to Lose Them* (Edinburgh: Q Press, 1978).
Gill, A., *Hessle Road: A Photographer's View of Hull's Trawling Days* (Beverley: Hutton Press, 1987).
Gill, A., *Superstitions: Folk Magic in Hull's Fishing Community* (Beverley: Hutton Press, 1993).
Gillett, E. and MacMahon K., *A History of Hull* (Hull: Hull University Press, 1980).
Halbwachs, M., *On Collective Memory* (London: University of Chicago Press, 1992).
Hewison, R., *The Heritage Industry: Britain in a Climate of Decline* (London: Methuen, 1987).
Holland-Martin, D., *Trawler Safety: Final Report of the Committee of Inquiry into Trawler Safety* (London: Committee of Inquiry into Trawler Safety, 1969).
Holm, P., Starkey D. J. and Thor, J. (eds), *The North Atlantic Fisheries 1100–1976: National Perspectives on a Common Resource*, Studia Atlantica: 1 (Esbjerg: NAFHA, 1996).
Hugill, S., *Sailortown*, 2nd edn (London: Routledge & Kegan Paul, 1967).
James, D., *Dona Maria's Story: Life History, Memory and Political Identity* (London: Duke University Press, 2000).
Jamieson, A., *Ebb Tide in the British Maritime Industries: Change and Adaptation 1918–1990* (Exeter: University of Exeter Press, 2003).
Johannesson, G., *Sympathy and Self-Interest: Norway and the Anglo-Icelandic Cod Wars* (Oslo: Institutt for Forvarsstudier, 2005).
Johannesson, G., *Troubled Waters: Cod War, Fishing Disputes and Britain's Fight for the Freedom of the High Seas*, Studia Atlantica: 11 (Hull: NAFHA, 2007).
Klein, J., *Samples from English Cultures* (London: Routledge & Kegan Paul, 1965).
Lazenby, C. and Lazenby, J., *Deep Sea Voices: Reflections of Women in our Fishing Communities* (Stroud: Tempus, 1999).
Le Sann, A., *A Livelihood from the Sea: Globalisation and Sustainable Fisheries Policies* (London: Intermediate Technologies Publications, 1998).
Leigh, M., *European Integration and the Common Fisheries Policy* (London: Croom Helm, 1983).
Neave, D. and Neave, S., *Hull*, Pevsner Architectural Guides (London: Yale University Press, 2010).

Parkes, B., *Trawlings of a Lifetime* (Penrith: Cumberland and Westmoreland Herald, 1991).

Perks, R. and Thomson, A., *The Oral History Reader*, 2nd edn (London: Routledge, 2006).

Piper, S., *The North Ships: The Life of a Trawlerman* (Newton Abbot: David & Charles, 1974).

Platten, D., *From the Arctic to Angola: The Adventures of a Hull Skipper* (Hull: Riverhead, 2010).

Relph, E., *Place and Placelessness* (London: Pion, 1976).

Roberts, E., *Women and Families: An Oral History, 1940–1970* (Oxford: Blackwell, 1995).

Robertson, I. (ed.), *Heritage from Below* (Farnham: Ashgate, 2012).

Robinson, R., *Trawling: The Rise and Fall of the British Trawl Fishery* (Exeter: University of Exeter Press, 1996).

Samuel, R., *Theatres of Memory. Vol. I* (London: Verso, 1994).

Saunders, R., *Yes to Europe* (Cambridge: Cambridge University Press, 2018).

Seamon, D., *A Geography of the Lifeworld* (London: Croom Helm, 1979).

Shackleton, M., *The Politics of Fishing in Britain and France* (Aldershot: Gower, 1986).

Smith, L., *Uses of Heritage* (London: Routledge, 2006).

Starkey, D. J. and Heidbrink, I. (eds), *A History of the North Atlantic Fisheries Vol. 2: From the 1850s to the Early Twenty-First Century* (Bremen: Verlag HM Hauschild, 2012).

Starkey, D. J. and Jamieson A. (eds), *Exploiting the Sea: Aspects of Britain's Maritime Economy Since 1870* (Exeter: University of Exeter, 1998).

Starkey, D. J., Reid, C. and Ashcroft, N. (eds), *England's Sea Fisheries: The Commercial Sea Fisheries of England and Wales since 1300* (London: Chatham, 2000).

Thompson, M., *Hull's Side-Fishing Trawling Fleet, 1946–86* (Beverley: Hutton Press, 1987).

Thompson, M., *Hull and Grimsby Stern Trawling Fleet 1961–1988* (Beverley: Hutton Press, 1988).

Thompson, M., *Fish Dock: The Story of St. Andrew's Dock Hull* (Beverley: Hutton Press, 1989).

Thompson, M., *The Marr Story, The History of a Fishing Family* (Beverley: J. Marr, 1995).

Thompson, M., *A Tribute to Hull's Fishing Industry* (Beverley: Hutton Press, 1996).

Thompson, P., *The Voice of the Past: Oral History*, 3rd edn (Oxford: Oxford University Press, 2000).

Thompson, P., Wailey, T. and Lummis, T., *Living the Fishing* (London: Routledge & Kegan Paul, 1983).

Thor, J., *British Trawlers and Iceland, 1919–1976* (Goteborg: University of Goteborg, 1995).

Triplow, N., *Pattie Slappers: Stories from the Golden Age of Hull's Food Processing Industry* (Immingham: Community Heritage Publishing, 2013).

Tuan, Y., *Topophilia: A Study of Environmental Perception, Attitudes and Values* (New York: Columbia University Press, 1974).

Tunbridge, J. E. and Ashworth, G. J., *Dissonant Heritage: The Management of the Past as a Resource in Conflict* (Chichester: Wiley, 1996).

Tunstall, J., *Fish: An Antiquated Industry*, Fabian Tract 360 (London: Fabian Society, 1968).

Tunstall, J., *The Fishermen: The Sociology of an Extreme Occupation*, 2nd edn (London: MacGibbon & Kee, 1969).

Walton, J., *Fish & Chips and the British Working Class, 1870–1940* (Leicester: Leicester University Press, 1992).

Williams, J., *Swinging the Lamp: Nautical Memories of a Lifetime at Sea* (Hull: Riverhead, 2013).

Wise, M., *The Common Fisheries Policy of the European Community* (London: Methuen, 1984).

Wright, P., *On Living in an Old Country: The National Past in Contemporary Britain* (London: Verso, 1985).

Wylie, J., *Landscape* (Abingdon: Routledge, 2007).

Essays in Books/Periodical Articles

Anderson, B. and Harrison, P., 'The Promise of Non-Representational Theory', in B. Anderson and P. Harrison (eds), *Taking-Place: Non-Representational Theories and Geography* (Surrey: Ashgate, 2010), 1–34.

Antonova, A. and Rieser A., 'Curating Collapse: Performing Maritime Cultural Heritage in Iceland's Museums and Tours', *Maritime Studies*, 18 (2019), 103–114.

Ashcroft, N., 'The Diminishing Commons: Politics, War and Territorial Waters in the Twentieth Century', in D. J. Starkey, C. Reid and N. Ashcroft (eds), *England's Sea Fisheries: The Commercial Sea Fisheries of England and Wales since 1300* (London: Chatham, 2000), 217–226.

Atkinson, D., 'The Heritage of Mundane Places', in B. Graham and P. Howard (eds), *The Ashgate Research Companion to Heritage and Identity* (Aldershot: Ashgate, 2008), 381–395.

Atkinson, D., Cooke, S. and Spooner, D., 'Tales from the Riverbank: Place-Marketing and Maritime Heritages', *International Journal of Heritage Studies*, 8:1 (2002), 25–40.

Balderstone, L., Milne, G. J. and Mulhearn, R., 'Memory and Place on the Liverpool Waterfront in the Mid-Twentieth Century', *Urban History*, 41:3 (2014), 478–496.

Benediktsson, K. and Waage E., 'Taskscapes at Sea: Trawler Fishing and the Experience of Fluid Nature', *Environment, Space, Place*, 7:2 (2015), 41–64.

Burton, A. and Ramster, J., 'Government, Science and the Fisheries in the Twentieth Century', in D. J. Starkey, C. Reid and N. Ashcroft (eds), *England's Sea Fisheries: The Commercial Sea Fisheries of England and Wales since 1300* (London: Chatham, 2000), 174–187.

Byrne, J., 'Hull, Fishing and the Life and Death of Trawlertown: Living the Spaces of a Trawling Port-City', in B. Beaven, K. Bell and R. James (eds), *Port Towns and Urban Cultures: International Histories of the Waterfront, c.1700–2000* (London: Palgrave Macmillan, 2016), 243–263.

Byrne, J. and Ombler, A., 'Memory on the Waterfront in Late Twentieth-Century Hull', in D. Starkey, D. Atkinson, B. McDonagh, S. McKeon and E. Salter (eds), *Hull: Culture, History, Place* (Liverpool: Liverpool University Press, 2017), 271–301.

Cadman, L., 'Nonrepresentational Theory/Nonrepresentational Geography', in R. Kitchin and N. Thrift (eds), *International Encyclopaedia of Human Geography* (Elsevier Science, 2009), 456–463.

Coopey, R. and Woodward, N., 'The British Economy in the 1970s: An Overview', in R. Coopey and N. Woodward (eds), *Britain in the 1970s: The Troubled Economy* (London: UCL Press, 1996), 1–33.

Crang, M. and Travlou, P., 'The City and Topologies of Memory', *Environment and Planning D: Society and Space*, 19 (2001), 161–177.

Crooke, E., 'The Politics of Community Heritage: Motivations, Authority and Control, *International Journal of Heritage Studies*, 16:1–2 (2010), 16–19.

Daunton, M., 'Jack Ashore: Seamen in Cardiff before 1914', *Welsh History Review*, 9 (1979), 176–203.

DeSilvey, C. and Edensor, T., 'Reckoning with Ruins', *Progress in Human* Geography, 37:4 (2012), 465–485.

Gibbs, D., 'Politics of Economic Development: Case of Mauritania Fishing Industry', *African Studies Review*, 27:4 (1984), 79–93.

Good, F., 'Voice, Ear and Text: Words, Meaning and Transcription', in R. Perks and A. Thomson (eds), *The Oral History Reader*, 2nd edn (Oxon: Routledge, 2006).

Graham, B., 'Heritage as Knowledge: Capital or Culture?' *Urban Studies*, 39: 5–6 (2002), 1003–1017.

Green, A., 'Individual Remembering and "Collective Memory": Theoretical Presuppositions and Contemporary Debates', *Oral History*, 32:2 (2004), 35–44.

Hagmark-Cooper, H., 'To be a Sailor's Wife: Ideals and Images of the Twentieth-Century Seafarer's Wife in the Aland Islands', in B. Beaven, K. Bell and R. James (eds), *Port Towns and Urban Cultures: International Histories of the Waterfront, c.1700–2000* (London: Palgrave Macmillan, 2016), 221–242.

Heidbrink, I., 'From Sail to Factory Freezer: Patterns of Technological Change', in D. J. Starkey and I. Heidbrink (eds), *A History of the North Atlantic Fisheries: From the 1850s to the Early Twenty-First Century* (Bremen: Verlag HM Hauschild, 2012), 58–71.

Hilling, D., 'Socio-economic Change in the Maritime Quarter: The Demise of Sailortown', in D. Pinder and S. Hussain (eds), *Revitalising the Waterfront* (London: Belhaven Press, 1988), 20–37.

Holm, P., 'The Modernisation of Fishing: The Scandinavian and the British Model', in R. Fischer, H. Hamre, P. Holm and J. Bruijn (eds), *The North Sea: Twelve Essays on Social History of Maritime Labour* (Stavanger: Stavanger Maritime Museum, 1992), 197–210.

Holm, P., 'An International Perspective on Britain's Fisheries in the New Millennium', in D. J. Starkey, C. Reid and N. Ashcroft (eds), *England's Sea Fisheries: The Commercial Sea Fisheries of England and Wales since 1300* (London: Chatham, 2000), 235–241.

Horobin, G. W., 'Community and Occupation in the Hull Fishing Industry', *British Journal of Sociology*, 8 (1957), 343–356.

Hubbard, J., 'Changing Regimes: Governments, Scientists and Fishermen and the Construction of Fishing Policies in the North Atlantic, 1850–2010', in D. J. Starkey and I. Heidbrink (eds), *A History of the North Atlantic Fisheries: From the 1850s to the Early Twenty-First Century* (Bremen: Verlag HM Hauschild, 2012), 129–168.

Inglis, F., 'Nation and Community: A Landscape and its Morality', *Sociological Review*, 25 (1977), 444–461.

Ingold, T., 'The Temporality of the Landscape', *World Archaeology*, 25:2 (1993), 152–174.

James, R., 'Cinema-going in a Port Town 1914–51: Film Booking Patterns at the Queens Cinema', *Urban History*, 40:2 (2013), 315–335.

Jamieson, A., 'An Inevitable Decline? Britain's Shipping and Shipbuilding Industries since 1930', in D. J. Starkey and A. Jamieson (eds), *Exploiting the Sea: Aspects of Britain's Maritime Economy Since 1870* (Exeter: University of Exeter, 1998), 79–92.

Johannesson, G., 'Life is Salt Fish: The Fisheries of the Mid-Atlantic Islands in the Twentieth Century', in D. J. Starkey and I. Heidbrink (eds), *A History of the North Atlantic Fisheries: From the 1850s to the Early Twenty-First Century* (Bremen: Verlag HM Hauschild, 2012), 277–288.

Jones, O., 'Geography, Memory and Non-Representational Geographies', *Geography Compass*, 5:12 (2011), 875–885.

Landzelius, M., 'Commemorative Dis(re)membering: Erasing Heritge, Spatializing Disinheritance', *Environment and Planning D: Society and Space* 21 (2003), 195–221.

Lazenby, C., 'Change and Adaptation in European Fishing Communities, 1970–1997', in P. Holm and D. J. Starkey (eds), *Technological Change in the North Atlantic Fisheries*, Studia Atlantica: 3 (Esbjerg: NAFHA, 1999), 187–204.

Lazenby, C. and Starkey, D. J., 'Altered Images: Representing the Trawling in the Late Twentieth Century', in D. J. Starkey, C. Reid and N. Ashcroft (eds), *The Commercial Sea Fisheries of England and Wales since 1300* (London: Chatham, 2000), 166–172.

Leeson, L., 'Our Land: Creative Approaches to the Redevelopment of London's Docklands', *International Journal of Heritage Studies*, 25:4 (2019), 365–379.

Lorimer, H., 'Cultural Geography: The Busyness of Being "More-Than-Representational"', Progress *in Human Geography*, 29 (2005), 551–559.

Mah, A., 'Memory, Uncertainty and Industrial Ruination: Walker Riverside, Newcastle upon Tyne', *International Journal of Urban and Regional Research*, 34:2 (2010), 398–413.

Mason, R. and Sayner, J., 'Bringing Museal Silence into Focus: Eight Ways of Thinking about Silence in Museums', *International Journal of Heritage Studies*, 25:1 (2019), 5–20.

Mumby-Croft, R. and Barnard, M., 'An Antiquated Relationship? Trawler Owners and Trawlermen, 1880–1980', in D. J. Starkey, C. Reid and N. Ashcroft (eds), *England's Sea Fisheries: The Commercial Sea Fisheries of England and Wales since 1300* (London: Chatham, 2000), 119–126.

Muzaini, H. and Minca, C., 'Rethinking Heritage, but "From Below"', in H. Muzaini and C. Minca (eds), *After Heritage: Critical Perspectives on Heritage from Below* (Cheltenham: Edward Elgar Publishing, 2018), 1–21.

Nora, P., 'Between Memory and History: Les Lieux de Memoire', *Representations*, 26 (1989), 7–27.

Orange, H., 'Industrial Archaeology: Its Place Within the Academic Discipline, the Public Realm and the Heritage Industry', *Industrial Archaeology Review*, XXX:2 (2008), 83–95.

Palmer, S., 'Ports', in M. Daunton (ed.), *The Urban History of Britain* (Cambridge: University Press, 2000), 133–150.

Portelli, A., 'What Makes Oral History Different', in R. Perks and A. Thomson (eds), *The Oral History Reader*, 2nd edn (Oxon: Routledge, 2006), 32–42.

Pred, A., 'Place as Historically Contingent Process: Structuration and the Time-Geography of Becoming Places', *Annals of the Association of American Geographers*, 74:2 (1984), 279–297.

Reid, C., 'From Trawler to Table: The Fish Trades since the Late-Nineteenth Century', in the Twentieth Century', in D. J. Starkey, C. Reid and N. Ashcroft (eds), *England's Sea Fisheries: The Commercial Sea Fisheries of England and Wales since 1300* (London: Chatham, 2000), 157–165.

Reid, C., 'Britain's Most Antiquated Industry: Mr Tunstall and the Fishing Industry', *International Journal of Maritime History*, 22:2 (2010), 171–198.

Robertson, I., 'Heritage from Below: Class, Social Protest and Resistance', in B. Graham and P. Howard (eds), *The Ashgate Research Companion to Heritage and Identity* (Aldershot: Ashgate, 2008), 143–158.

Robinson, R., 'The Line and Trawl Fisheries in the Age of Sail', in D. J. Starkey, C. Reid and N. Ashcroft (eds), *England's Sea Fisheries: The Commercial Sea Fisheries of* England *and Wales since 1300* (London: Chatham, 2000), 72–80.

Robinson, R., 'Steam Power and Distant-Water Trawling', in D. J. Starkey, C. Reid and N. Ashcroft (eds), *England's Sea Fisheries: The Commercial Sea Fisheries of England and Wales since 1300* (London: Chatham, 2000), 206–216.

Robinson, R., 'Hook, Line and Sinker: Fishing History – Where Have We Been, Where Are We Now and Where Are We Going?' *Mariner's Mirror*, 97:1 (2011), 167–179.

Rose, M., 'Landscape and Labyrinths', *Geoforum*, 33 (2002), 455–467.

Scannell, L. and Gifford, R., 'The Experienced Psychological Benefits of Place Attachment', *Journal of Environmental Psychology*, 51 (2017), 256–269.

Schofield, J., 'Heritage Expertise and the Everyday: Citizens and Authority in the Twenty-first Century', in J. Schofield (ed.), *Who Needs Experts: Counter-Mapping Cultural Heritage* (Farnham: Ashgate, 2014), 1–11.

Starkey, D. J., 'Distant-Water Trawlerman: William Oliver, 1884–1959', in D. J. Starkey, D. Atkinson, B. McDonagh, S. McKeon and E. Slater (eds), *Hull: Culture, History, Place* (Liverpool: Liverpool University Press, 2017), 207–237.

Symes, D., 'The European Community's Common Fisheries Policy', *Ocean & Coastal Management*, 35:2–3 (1997), 137–155.

Tenold, S., 'The Declining Role of Western Europe in Shipping and Shipbuilding, 1900–2000', in N. Petersson, S. Tenold and N. White (eds), *Shipping and Globalization in the Post-War Era* (Open Access: Palgrave Macmillan, 2019), 9–33.

Thompson, P., 'Women in the Fishing: The Roots of Power between the Sexes', *Comparative Studies in Society and History*, 27:1 (1985), 3–32.

Thrift, N., 'Dead Geographies – and How to Make them Live', *Environment and Planning D: Society and Space*, 18 (2000), 411–432.

Till, K., 'Places of Memory', in J. Agnew, K. Mitchell and G. Toal (eds), *A Companion to Political Geography* (Oxford: Blackwell, 2003), 289–301.

Van Huis, I., Kaasik-Krogerus, S., Lahdesmaki, T. and Ellena, L., 'Introduction: Europe, Heritage and Memory – Dissonant Encounters and Explorations', in T. Lahdesmaki, L. Passerini, S. Kaasik-Krogerus and I. van Huis (eds), *Dissonant Heritages and Memories in Contemporary Europe* (Palgrave Macmillan, 2019), 1–21.

Yang, L, Dupre, K. and Jin, X., 'A Systematic Review of Literature on Contested Heritage', *Current Issues in Tourism*, 24:4 (2021), 442–465.

Yow, V., 'Do I Like them Too Much? Effects of the Oral History Interview on the Interviewer and Vice-Versa', in R. Perks and A. Thomson (eds), *The Oral History Reader*, 2nd edn (Oxon: Routledge, 2006), 54–72.

Walton, J., 'Fishing Communities, 1850–1950', in D. J. Starkey, C. Reid and N. Ashcroft (eds), *The Commercial Sea Fisheries of England and Wales since 1300* (London: Chatham, 2000), 127–137.

Wetherell, M., Smith, L. and Campbell, G., 'Introduction: Affective Heritage Practices', in L. Smith, M. Wetherell and G. Campbell (eds), *Emotion, Affective Practices and The Past in the Present* (Abingdon: Routledge, 2018), 1–21.

Wilcox, M., 'Beyond the North Atlantic', in D. J. Starkey and I. Heidbrink (eds), *A History of the North Atlantic Fisheries: From the 1850s to the Early Twenty-First Century* (Bremen: Verlag HM Hauschild, 2012), 306–326.

Wilcox, M., '"To Save the Industry from Complete Ruin": Crisis and Response in British Fishing 1945–1951', *Business History* (2019), 353–377.

Wolferstan, S., 'An Ethnography of a "Humble Expert": Experiencing Faro', in J. Schofield (ed.), *Who Needs Experts: Counter-Mapping Cultural Heritage* (Farnham: Ashgate, 2014), 43–53.

Ziller, J., 'The European Union and the Territorial Scope of European Territories', *Victoria University of Wellington Law Review*, 38:1 (2006), 51–63.

Unpublished Sources

Bell, K., 'Port-city Afterlives: The Cultural Function of Ghosts in Nineteenth-century Portsmouth', paper presented at Port City Lives Conference, Liverpool University, 29–30 June 2012.

Loader, J., 'The Decline of Hull's Fishing Industry in the Last Quarter of the 20th Century', MA Thesis, University of Hull, 2010.

Moon, L., 'Sailorhoods: Sailortown and Sailors in the port of Portsmouth circa 1850–1900', PhD Thesis, University of Portsmouth, 2015.

Stephens, T., 'Why Does Sudden Loss Continue to Feature So Much in the Current Pastoral Work of the Fishermen's Mission in Hull', Undergraduate Dissertation, William Booth College and Havering College, 2012.

Film/DVD/CD

Conolly, J and Sumner, P., *Trawlertown: The Singing of the Fishing* (Fellside, 2000) [CD].

Gill, A., *Three-Day Millionaires: Trawlermen Home Between Trips* (Hull: Dovedale Studios, 2001) [DVD].

Trawlertowns – Hull and Grimsby (Forest Edge, 1993) [VHS].

Trawling – Your Future (Antony Barrier Productions, 1976) [documentary film: British Film Institute Ref: C-64335].

Online Sources

Change.org, https://www.change.org/p/hull-city-council-we-the-hull-fishermens-history-pages-on-facebook-with-4430-members-call-upon-hull-city-council-to-reclaim-lordline-building-bullnose-bostons-building-and-the-surrounding-area-back-to-their-control-and-provide-a-memorial-for-the-lost (accessed 11 February 2015).

Dukes Wood Oil Museum, www.dukeswoodoilmuseum.co.uk/sea-gem.htm (accessed 30 July 2014).

Falkland: Our Islands, Our History, www.falklands.gov.fk/assets/OurIslandsOurHistory.pdf (accessed 11 August 2014).

Handling and Processing Blue Whiting, Torry Research Station, www.fao.org/wairdocs/tan/x5952e/x5952e01.htm (accessed 14 July 2014).

Hull Bullnose Heritage Group, www.hullbullnoseheritagegroup.co.uk/?page_id=11 (accessed 28 July 2020).

Hull Daily Mail, www.hulldailymail.co.uk/Pictures-new-memorial-Hull-s-lost-fishermen/story-259511-detail/story.html (accessed 12 February 2015); www.hulldailymail.co.uk/Ex-fishermen-want-bullnose-rethink-St-Andrew-s-story-19609477-detail/story.html (accessed 12 February 2015); www.hulldailymail.co.uk/New-plan-Bullnose-memorial-Hull-s-lost-trawlermen/story-22084420-detail/story.html (accessed 12 February 2015).

MV Miranda, http://nigelhadley.org.uk/html (accessed on 17 May 2014).

STAND, http://www.hullfishingheritage.org.uk/lost-trawlermens-day/ (accessed 21 July 2020); https://www.hullfishingheritage.org.uk/memorial-heritage-garden/ (accessed 21 November 2020).

Index